Gordon G. Wallace, Simon E. Moulton, Robert M. I. Kapsa, and Michael J. Higgins

Organic Bionics

Related Titles

Mano, J. F. (ed.)

Biomimetic Approaches for Biomaterials Development

2012
Hardcover
ISBN: 978-3-527-32916-8

Fratzl, P., Harrington, M. J.

Introduction to Biological Materials Science

2013
Hardcover: ISBN: 978-3-527-32971-7
Softcover: ISBN: 978-3-527-32940-3

Öchsner, A., Ahmed, W. (eds.)

Biomechanics of Hard Tissues

Modeling, Testing, and Materials

2010
Hardcover
ISBN: 978-3-527-32431-6

Carpi, F., Smela, E. (eds.)

Biomedical Applications of Electroactive Polymer Actuators

2009
Hardcover
ISBN: 978-0-470-77305-5

Hadziioannou, G., Malliaras, G. G. (eds.)

Semiconducting Polymers
Chemistry, Physics and Engineering

2007
Hardcover
ISBN: 978-3-527-31271-9

Gordon G. Wallace, Simon E. Moulton, Robert M. I. Kapsa, and Michael J. Higgins

Organic Bionics

WILEY-VCH

WILEY-VCH Verlag GmbH & Co. KGaA

The Authors

Prof. Gordon G. Wallace
AIIM Facility
University of Wollongong
ARC Centre of Excellence for
Electromaterials Science/
Intelligent Polymer
Research Institute
Wollongong
New South Wales 2522
Australia

Dr. Simon E. Moulton
AIIM Facility
University of Wollongong
ARC Centre of Excellence for
Electromaterials Science/
Intelligent Polymer
Research Institute
Wollongong
New South Wales 2522
Australia

Prof. Robert M. I. Kapsa
The University of Wollongong
St Vincent's Hospital
41 Victoria Pde, Fitzroy
Victoria 3065
Australia

Dr. Michael J. Higgins
AIIM Facility
University of Wollongong
ARC Centre of Excellence for
Electromaterials Science/
Intelligent Polymer
Research Institute
Wollongong
New South Wales 2522
Australia

Library of Congress Card No.: applied for

**British Library Cataloguing-in-Publication
Data**
A catalogue record for this book is available
from the British Library.

**Bibliographic information published by the
Deutsche Nationalbibliothek**
The Deutsche Nationalbibliothek
lists this publication in the Deutsche
Nationalbibliografie; detailed bibliographic
data are available on the Internet at
<http://dnb.d-nb.de>.

© 2012 Wiley-VCH Verlag & Co. KGaA,
Boschstr. 12, 69469 Weinheim, Germany

Print ISBN: 978-3-527-32882-6
ePDF ISBN: 978-3-527-64605-0
ePub ISBN: 978-3-527-64604-3
Mobi ISBN: 978-3-527-64603-6
oBook ISBN: 978-3-527-64602-9

Cover Design   Formgeber, Eppelheim
Typesetting   Laserwords Private Limited,
Chennai, India
Printing and Binding Strauss GmbH,
Mörlenbach

Printed in the Federal Republic of Germany
Printed on acid-free paper

Contents

Foreword by Professor Graeme Clark

The advent of the bionic ear relied on the identification, optimization and development of appropriate materials that would provide an effective biological – electronic interface [1]. It also needed the development of appropriate electronics and communications (software) strategies and most critically the effective integration of a range of skills and their accompanying personalities to form a multidisciplinary research and clinical team that could ensure success.

The discovery of organic conductors (by MacDiarmid, Shirakawa and Heeger [2] in 1976) provided a new set of materials with properties that have excited those involved in bionics research. Their stimuli-responsive, multifunctional nature provides an unprecedented communications conduit from the cold hard world of electronics to the warm but dynamic world of biology providing new opportunities to bridge the electrode-cellular interface [3].

When configured appropriately these soft organic conductors can even provide some of the electronic components usually associated with hard materials such as silicon. In addition, the multifunctionality of these structures provides new opportunities but also new challenges.

The rational design of new bionic interfaces containing them is at the moment challenging.

The materials are usually not amenable to conventional processing and device fabrication methodologies, and "seeing" how they perform in simulated operational environments (fluids) – with the need for spatial information (in the nanodomain) to provide topographical, chemical, mechanical and biological maps of surfaces – is challenging. However, some significant steps forward have been achieved in this direction using biological-atomic force microscopy Bio-AFM.

Material characterization is obviously important to provide a rational approach during each stage of device fabrication and development. But is it just as critical for acquiring as much information on these materials/material configurations as is humanely possible so that there is sufficient knowledge to pass these materials through the rigorous screening processes on the path to FDA approval, ethics and societal acceptance.

Success will depend not only on overcoming these challenges but also in building effective, integrated multidisciplinary research and clinical teams to implement the scientific breakthroughs in practical devices.

In parallel, ethical issues and the most appropriate ways to bring along the general community (to ensure social acceptance) must be addressed.

References

1. Clark, G. (2000) *Sounds from Silence*, Allen & Unwin.
2. Nobelprize.org. The Nobel Prize in Chemistry 2000 (accessed 12 January 2012) *http://www.nobelprize.org/nobel_prizes/chemistry/laureates/2000/*.
3. Wallace, G.G., Moulton, S.E., and Clark, G.M. (2009) Electrode-Cellular Interface. *Science*, **324** (5925), 185–186.

Acknowledgments

The authors are indebted to a number of significant figures who inspired our venture into this fascinating field of research.

Prof. Graeme Clark inspired us in that impossible research goals could be achieved, in that advances in any field even ones as complex as bionics can be forged with the development of the appropriate multidisciplinary research team. Prof. Alan MacDiarmid who in the early years encouraged us to explore the bioapplication of organic conductors was an enduring source of inspiration and encouragement.

We are also indebted to numerous coworkers in our own laboratories and from other laboratories around the world with whom we have worked shoulder to shoulder on numerous aspects of organic bionics. In particular, we thank those associated with the ARC Centre of Excellence for Electromaterials science.

GGW is particularly indebted to the coauthors of *Conductive Electroactive Polymers* 3rd Edition (Spinks, Kane-Maguire and Teasdale). Many of the discussions emanating in the development of that monograph sowed the seeds for this text on organic bionics.

Each of the coauthors thank their families for making seemingly endless tasks such as the development of this monograph possible and so we thank Vicky, Jordan and Eileen Wallace, Louise, Aleida and Liam Moulton, Lorraine Shine, Eve and Georgia Higgins, and Anita Quigley.

Finally, we thank the global scientific research community, it is a special family to which we are all privileged to belong.

Professor Gordon G. Wallace

Dr. Simon E. Moulton

Professor Robert M.I. Kapsa

Dr. Michael J. Higgins

1
Medical Bionics

The term *"bionics"* is synonymous with *"biomimetics"* and in this context refers to the integration of human-engineered devices to take advantage of functional mechanisms/structures resident in Nature. In this book, we refer to the field of bionics, and in particular medical bionics, as that involved with the development of devices that enable the effective integration of biology (Nature) and electronics to achieve a targeted functional outcome.

Since the early experiments of Luigi Galvani and Alessandro Volta (see insets), the use of electrical conductors to transmit charge into and out of biological systems to affect biological processes has been the source of great scientific interest. This has inspired many to explore the possible use of electrical stimulation in promoting positive health outcomes. Some of the earliest examples of using electrical stimulation in a controlled manner to achieve specific clinical outcomes were developed by Guillaume-Benjamin-Amand Duchenne (see inset) (Figure 1.1). Duchenne's interests in physiognomic esthetics of facial expression led to the definition of neural conduction pathways. During this important period in the history of science, Duchenne developed nerve conduction tests using electrical stimulation and performed pioneering studies of the manner in which nerve lesions could be diagnosed and possibly treated.

To date, medical bionic devices have been largely targeted toward the primary "excitable cell" systems, muscle, and nerve, whose functions are inherently capable of being modulated by electrical stimulation. There have also been numerous studies of the use of electrical stimulation for bone regrowth and wound healing. The effects of electrical stimulation are thought to be promoted through the induced movement of positive and negative charged ions in opposite directions (polarization) across cells and tissues that activates sensory or motor functions [2].

Landmark developments such as the artificial heart in 1957 (Kolff and Akutsu) [3, 4] the external (1956) [5] and then implantable (1958) [6] cardiac pacemaker; the artificial vision system (1978) [7]; the cochlear implant (1978) [8, 9] deep brain stimulation (DBS) electrodes (1987) [10]; and, more recently, electroprosthetic limbs [11] are now being used along with a broad spectrum of parallel developmental projects that aim to alleviate human afflictions.

Organic Bionics, First Edition. Gordon G. Wallace, Simon E. Moulton, Robert M.I. Kapsa, and Michael J. Higgins.
© 2012 Wiley-VCH Verlag GmbH & Co. KGaA. Published 2012 by Wiley-VCH Verlag GmbH & Co. KGaA.

Luigi Galvani was born 9 September, 1737, in Bologna, Italy. He was educated at Bologna's medical school and in 1762 was appointed as a public lecturer in Anatomy. In 1791, he published the "Commentary on the Effect of Electricity on Muscular Motion" in the seventh volume of the Proceedings of the Institute of Sciences at Bologna.

Galvani was dissecting a frog on a table near a wheel that generated static electricity, which he had been using for a physics experiment. As Galvani put the scalpel to the sciatic nerve, which innervates the muscles in the frog's legs, a spark was discharged from the wheel and the frog's legs jerked. The static electricity was picked up by the scalpel and passed to the nerve. Galvani conducted other experiments in this area, one of which used the electricity from a thunderstorm to enable a frog's legs to appear to "dance."

Alessandro Volta was born 18 February 1745, in Como, Italy. He was appointed as a Professor of Physics at the Royal School of Como. Around 1790, Volta took an interest in the "animal electricity" discovered by Galvani. Volta built on this experimental observation, replacing the frog's legs with a more traditional electrolyte to create the world's first galvanic cell and subsequently the battery. Volta also had an interest in things pertaining to medical bionics. He recorded an  experiment wherein he placed metal rods attached to an active electrode circuit into his ears and reported a sound similar to boiling water.

Guillaume-Benjamin-Amand Duchenne (de Boulogne) was born 17 September 1806, in Boulogne-sur-Mer, France. Duchenne was a French neurologist who followed on from Galvani's research, making seminal contributions to the clinical area of muscle electrophysiology.

After practicing as a physician in Boulogne for four years, in 1835, Duchenne began experiments on the potential of subcutaneous "electrotherapy" to treat various muscle conditions.

Duchenne returned to Paris in 1842, where his research yielded a non-invasive electrical technique for muscle stimulation that involved the delivery of a localized faradaic shock to the skin ("faradization"). He articulated these theories in his work, *On Localized Electrization and its Application to Pathology and Therapy*, first published during 1855 [1].

Graeme Milborne Clark was born 16 August, 1935, in Camden, Australia. He studied medicine at the University of Sydney and continued his studies in general surgery at the Royal College of Surgeons, Edinburgh.

In 1927, at age 22, his father, a pharmacist, noticed a decrease in his hearing and was fitted with a hearing aid in 1945. Graeme's fascination with medical science and his father's plight led him on a journey to create the multielectrode bionic ear implant, which has now been implanted in more than 100 000 adults and children around the world.

Clark's story [1] is an inspiration not only in the science behind the invention but also in the tenacity and determination shown in building a multidisciplinary research team while "securing" funding and traversing adversity.

Alan MacDiarmid was born in Masterton, New Zealand, on 14 April, 1927. He developed an interest in chemistry at around age 10 and taught himself from one of his father's textbooks. Formally educated at Hutt Valley High School and Victoria University in Wellington, New Zealand, MacDiarmid obtained his first Ph.D. degree from the University of Wisconsin-Madison in 1953. A second Ph.D. degree was awarded to him from Cambridge, United Kingdom, in 1955.

MacDiarmid went on to discover organic conducting polymers (OCPs) and, together with his colleagues, Hideki Shirakawa and Alan Heeger, was awarded the Nobel Prize in the year 2000. MacDiarmid continued to pioneer applications for OCPs until he passed away in 2006. He was an inspiration to all researchers young and old, mentoring all with his view on success published in his Nobel Prize autobiography: "Success is knowing that you have done your best and have exploited your God-given or gene-given abilities to the next maximum extent. More than this, no one can do."

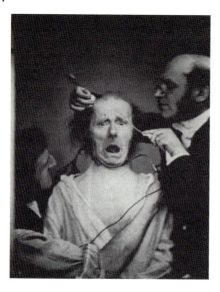

Figure 1.1 Demonstration of the mechanics of facial expression using electrical stimulation. The test subject, a cobbler by trade and a patient of Duchenne's, is "faradized" by Duchenne (right) and his assistant (left). The stimulation was applied to the cobbler's mimetic (facial) muscles and caused a change in his facial expression.

Advances in medical bionics technology are dependent on eliciting precise control of the electrical energy to deliver beneficial health outcomes. The advent of carbon-based organic conductors, through the pioneering works of Alan MacDiarmid, Alan Heeger, and Hideki Shirakawa (see inset), now provides the platform for unprecedented possibilities by which electrical energy can be used to modulate the function of medical devices.

There are three main application paradigms where the use of organic conductors as electrodes or electrode arrays in medical bionic devices could be beneficial – where the organic polymer is used to functionally stimulate the target tissue, stimulate a regenerative event in the target tissue, or stimulate communication between the nervous system and an electronically driven prosthesis.

1.1
Medical Bionic Devices

1.1.1
Electrodes and Electrode Arrays

Electrodes may be surgically implanted within the body to record (e.g., brain activity) and/or stimulate (i.e., cardiac pacing, DBS, and bone growth) function. The specific material requirements for these electrodes will differ markedly in accordance with their proposed interaction with the tissues that they are intended to stimulate. In all instances, the conducting surface needs to be able to facilitate control of the electron flux, without the promotion of adverse effects on the implant's tissue environment. Furthermore, for some applications, the conducting surface needs to be in direct contact with the surrounding tissue, while in other applications, no direct contact is required. It is in the modulation of the interaction between the target tissue and conducting surface that nonmetallic conductors, such as OCPs and/or conducting carbon materials, are able to add value toward further optimization of electrodes [12].

Neurophysiologists have used sharpened wire metal (e.g., tungsten) electrodes for over 50 years to study brain function [13], and recently, neural probes have been fashioned from silicon [14], ceramic [15], and flexible substrates [16, 17]. However, in all cases, the surfaces from which the bioelectrical impulses are transferred between tissue (e.g., neural) and electronic circuitry remain predominantly metallic. Traditionally, the materials of choice for implantable electrodes have been based on inert metals.

The type of metal, its area of exposure, and the texture of the metal surface determine the properties of the electrodes and therefore the area of application. Activated Ir oxide has been shown to have excellent charge transfer properties ($3000 \, C \, \mu m^{-2}$), making it the material of choice for microelectrodes, but the surface is chemically unstable [18].

To enhance the metal electrode sensitivity or increase the electrode's capacity to conduct charge for use in stimulation or sensory monitoring, the impedance must be lowered [18]. This step generally involves increasing the geometric surface area of the electrode tip, often associated with a concomitant loss in resolution or either the stimulatory or recording process. Larger area electrodes cause increased tissue damage during insertion [19] or, if required, during removal.

Platinum (Pt) has been employed in cochlear implant electrodes as well as in other functional stimulation electrodes, such as pacemakers, early DBS electrodes, and vision stimulator applications. Gold, iridium (Ir) oxide [20], and alloys of Pt and Ir have likewise been incorporated into bioelectrodes for a wide variety of applications [21–23].

1.1.1.1 Bionic Hearing

The earliest report of electrical stimulation of the nervous system to elicit auditory sensations has somewhat inauspicious, yet subtly elegant, origins. In 1779, Italian physicist Volta made several discoveries of great importance relating to electricity, including the electrochemical battery. During this period (1800s), Volta argued with Galvani over Galvani's "animal electricity" postulation and famously performed an experiment on himself in which he attached batteries to two metal rods, each of which he then inserted into his ears. On closing the circuit, he described receiving a sharp jolt to the head, followed by "a kind of crackling, jerking, or bubbling as if some dough or thick stuff was boiling" [24].

Almost two centuries later, after it was reported that a deaf patient undergoing neurosurgery heard a noise in response to electrical stimulation of the auditory nerve [25], significant scientific interest focused on the possibility that hearing could be restored by electrical stimulation. This led to the first implant for stimulation of the saccular nerve in milestone experiments by Djourno and Eyries [26] who used an induction coil that connected an electrode in the inner ear with an electrode in the temporal muscle. In this elegant experiment, a second external induction coil was used to transmit signals (pulse rates between 100 and 1000 Hz) generated from a microphone. The subject was then able to accurately identify changes in signal pulse stimulus frequency, and reported a sensation of background noise as well as specific sounds resembling "the chirping of a cricket and the spinning of a roulette wheel."

This landmark finding was the first instance in which it was specifically demonstrated that hearing could be restored in bilaterally deaf subjects by stimulation of the auditory neural pathways. Subsequent experimentation saw the parallel development of single-channel electrodes from two groups [27, 28] and their application to cochlear nerve stimulation. Resulting from this, a cochlear stimulation system developed by House [29], involved a single gold electrode that was initially adopted for production by 3M and was subsequently implanted in some 1000 people worldwide. As this device was undergoing U.S. Food and Drug Administration (FDA) approval for use in humans, Graeme Clark and coworkers [30] continued their work on the development of a multichannel implant. The multichannel cochlear implant, first implanted in a human subject in 1978, was marketed world-wide by Cochlear Pty Ltd. and has since restored hearing in more than 100 000 people worldwide. Figure 1.2 shows the integration of the main components of the cochlear implant. During its operation, sound picked up through an external microphone is encoded through a speech processor and delivered to a receiver stimulator. Resulting electrical impulses are deliv-ered in appropriate spatiotemporal patterns to electrodes within the cochlea that stimulates the auditory neurons within the brain. The stimulatory system utilizes an array of 22 platinum microelectrodes, individually ad-dressed through connections integrated throughout a silicon-rubber-based housing. The receiver electronics are housed in a titanium casing. The implantable component of the device was initially the receiver/stimulator package, but, recently, Cochlear developed a totally implantable cochlear implant (TICI) device in which the entire transmitter is implanted within (Figure 1.3).

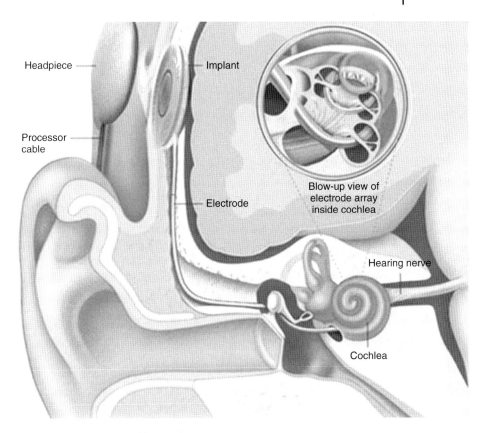

Figure 1.2 Bionic ear: cochlear implant [30]. (Image obtained with permission from *http://www.advancedbionics.com.*)

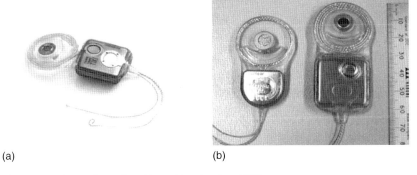

(a)   (b)

Figure 1.3 The totally implantable cochlear implant (TICI) system from Cochlear (a) is all internally implanted, including the power source and a subcutaneous microphone. (b) The TICI system (right device) compared to the 22-electrode "standard" cochlear implant (left device).

1.1.1.2 **Bionic Vision**

In 1755, some 36 years before Galvani communicated his "animal electricity" theory, the French chemist and physician Charles le Roy used a Leyden jar capacitor to noninvasively apply transcranial electrical stimulation to the head of a 21-year-old blind patient in the hope that this would restore the man's vision [31–33]. In what may be the first known example of a therapeutic bioelectrode system, three electrodes were anatomically aligned, two juxtaposed to the supraorbital ridges (Q in Figure 1.4) and one on the occupant (P in Figure 1.4), to stimulate the primary visual nerve tracts [32, 33]. The electrodes were connected by brass or iron wire and wound two to three times around the head; a wire leading to the right leg and a Leyden jar capacitor completed the circuit. While the patient reported sensation of phosphenes (percepts of light) after being shocked, he nevertheless remained blind despite several subsequent applications of the procedure.

Despite these early heroic forays into attempting to restore vision using electrical stimulation, no successful strategies, let alone therapies, emerged until the mid-twentieth century. Nevertheless, in terms of medical bionic developments, it is within the field of vision bionics that the most intensive research activity has occurred.

In the present day, significant advancement in knowledge on the biological and engineering prerequisites for the restoration of vision has resulted in the evolution of multiple approaches that can provide significantly greater scope for success than has been attainable in the past. Collectively, these

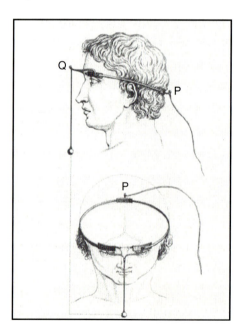

Figure 1.4 Charles le Roy's electrical stimulation experiment to restore vision.

approaches provide unprecedented choice as to where along the optic neural circuitry an electrode may be applied to effect visual percepts in blind subjects. Such choices currently being explored include stimulation of existing functional neurons in the retina [34–36] or direct stimulation of the optic nerve [23, 37, 38]. In cases where these regions are functionally compromised, the neurons within the visual cortex of the central nervous system (CNS) [39, 40] provide another target of interest.

Current strategies for restoring lost vision by electrical stimulation of visual neural pathways within the visual cortex involve more complex electrode configurations ranging from a four-contact single cuff [37], a relatively thrifty (up to 16 electrodes) multielectrode array (MEA) [23] for optic nerve stimulation, and "higher order" silicon-based MEAs (initially an 81-electrode array) [41] for cortical stimulation and cover several target foci along the neuroptic pathway.

These well-established approaches are bolstered by ongoing developmental strategies, such as suprachoroidal transretinal stimulation with electrodes between the sclera and the vitreous chamber [42], that have yet to be applied to human subjects.

Stimulation at the Retina   In some cases of blindness, the neural deficit occurs at the start of the neuroptic pathway. Conditions such as retinitis pigmentosa and age-related macular degeneration affect the retina, but the cones and rods that transfer visual signaling to the CNS are left intact. These conditions require (arguably) the least neuroinvasive prosthetic interventions and, as such, bionic prosthetics for such conditions focus on electrical stimulation at the retina.

Numerous animal studies indicating that the visual cortex could be stimulated in the absence of photoreceptors and characterizing the currents required to stimulate retinal ganglia [43–45] inspired scientific interest in the possibility that epiretinal (i.e., on the retina) electrical stimulation could restore vision in conditions where the point of neurological damage or deficit was the retina itself. This was confirmed in subsequent human studies, which demonstrated that stimulation of the retinal surface of blind subjects was able to generate phosphenes [16]. These phosphenes were able to be spatially characterized, such that the visual perception was sufficient for the blind subject to perceive movement after the implant. Follow-up experiments showed vast improvements on the earlier results, allowing the individual to perceive shapes [47]. This landmark finding spurred substantial research activity in this area, during which time much improvement on the initial application of the concept was achieved [48, 49].

There are five major initiatives to deliver retinal prostheses to blind humans. Three of these programs (Second Sight Argus, EPI-RET, and Intelligent Medical Implants) utilize epiretinal approaches, while the remaining two (Boston Retinal Implant and Retina Implant AG) adopt subretinal approaches. All of these systems generally utilize thin-film MEAs, with the electrodes made from platinum, platinum/iridium, iridium oxide, or titanium nitride [50].

One system, developed as part of the United States' Department of Energy Artificial Retina Project, consisted of a small camera and transmitter mounted in spectacles that communicated visual fields to the brain via an implanted receiver electrically stimulating a 16-channel MEA (initially) attached to the retina. The system was powered by a battery pack worn on the belt via a wireless microprocessor (Figure 1.5).

In 2002, the Argus I (a 16-channel MEA) was implanted onto the retina of a 77-year-old blind man as part of a microelectronic artificial vision system (Figure 1.6a,b). As a result, the man, who had been blind for the past 50 years, was able to perceive motion, distinguish objects, and distinguish patterns of light and dark [52]. Between 2002 and 2009, a further five people were implanted as part of the first phase of clinical trials with the Argus I device. Since 2009, some 30 legally blind subjects have been implanted with a second-generation device utilizing a 60-electrode MEA (Argus II) [51] as part of phase II/III clinical trials sponsored by Second Sight Medical Products Inc. The ultimate goal is to design a device that contains hundreds to more than a thousand microelectrodes (Figure 1.7). If successful, this device will help restore vision to people blinded by retinal disease, and enable reading, unaided mobility, and facial recognition.

The Boston Retinal Implant Project is focused on developing a system for restoration of vision and is based on a visual prosthesis that receives two types of input, namely, (i) information about the visual scene ("visual signal") and (ii) power and data to run the electronics and stimulate the retina in order to generate the corresponding visual image (Figure 1.6c,d). This system uses wireless communication utilizing radiofrequency (RF) to transmit data and power to receiving coils located on the prosthesis. Electrical current passing from individual electrodes (implanted within the retina) stimulate cells in the appropriate areas of the retina corresponding to the features in the visual scene. The results of initial trials were encouraging although the prosthesis has only been trialed on six human patients in a controlled surgical environment, with each trial lasting several hours. These tests showed that the stimulating electrode placed over the retina was able to stimulate enough retinal nerves in the patients who had been legally blind for decades, so they were able to see relatively small spots of light (i.e., like a "pea" as if viewed at arm's length). Occasionally, they were able to distinguish two spots of light from one another and "see" a line.

Stimulation of the Optic Nerve Emanating from the Université Catholique de Louvain (Brussels, Belgium), a four-contact cuff electrode was used by Claude Veraart for stimulation of the optic nerve and implanted into a 52-year-old woman who had lost her vision as a result of retinitis pigmentosa

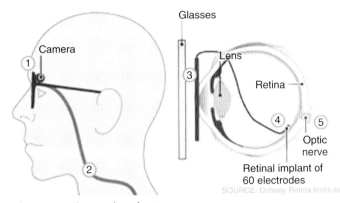

- 1: Camera on glasses views image
- 2: Signals are sent to handheld device
- 3: Processed information is sent back to glasses and wirelessly transmitted to receiver under the surface of the eye
- 4: Receiver sends information to electrodes in retinal implant
- 5: Electrodes stimulate retina to send information to the brain

(a)

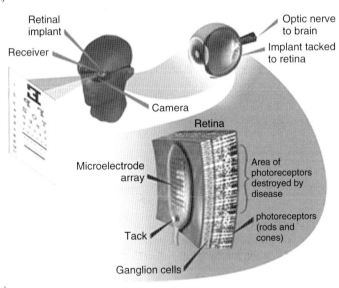

(b)   Adapted, with permission, from *IEEE Engineering in Medicine and Biology* 24:15 (2005).

Figure 1.5 US Department of Energy Artificial Vision System: (a) schematic of the system components and (b) schematic of the system's function to stimulate neurons in the retina. (Adapted from Chader and coworkers [51].)

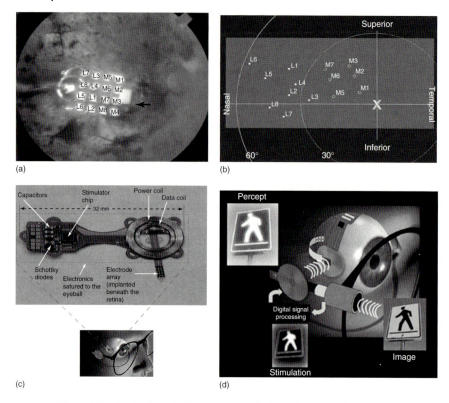

(a)   (b)   (c)   (d)

Figure 1.6 The implanted electrode array with electrode positions is shown two weeks after surgical implantation on the retina (a). The black arrow indicates a clinical reference point. Representation of the percepts elicited by stimulation of the electrodes relative to the implanted subject's visual field (b). Not all electrodes are included because of issues with the threshold current required to elicit a response at these initial stages of the implant's application. Picture (c) and schematic (d) of the Boston Retinal Implant System for Restoration of Vision. (Image adapted with permission from Humayun and coworkers [52].)

[53]. In this application, the intracranially implanted electrode was connected through the skull and skin to an extracranial stimulator. Phosphenes were generated by mono- and bipolar electrical stimulation paradigms. This continues to be developed as the Microsystems Visual Prosthesis (MiViP) system (Figure 1.8) by Veraart and colleagues [54] and is the only system that has reported any color modulations associated with the stimulation.

In another approach to optic nerve stimulation, the C-Sight program [23] has adopted the use of penetrating platinum/iridium MEAs by which to stimulate the optic nerve. In this adaptation, the vision capture device will be implanted into the subject's eye, rather than using spectacles with an inbuilt camera/lens (Figure 1.9).

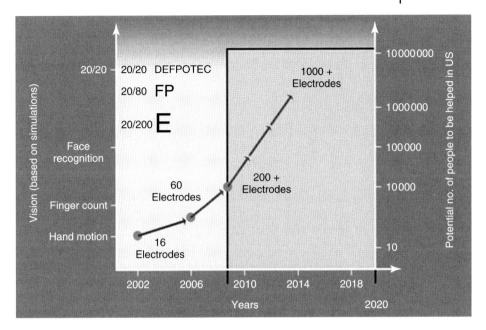

Figure 1.7 Milestone map for the US DOE's Artificial
Retina Program. (Adapted with permission from the
Department of Energy newsletter, 5 January 2008; taken from
Chader and coworkers [51].)

Stimulation of the Visual Cortex After research identified that stimulation
of the calcarine cortex produced visual sensations in humans with nonfunc-
tional eyes in 1967, Brindley and Lewin [41] implanted an 81-electrode array
(platinum encased in silicon) of subdural nonpenetrating electrodes over the
occipital cortex of a 52-year-old woman who lost her vision because of retinal
detachment (Figure 1.10). In these experiments, which laid the foundation
for cortical electrical stimulation to restore lost vision, approximately half of
the implanted electrodes were functional and produced phosphenes within
the subject's visual field.

Indeed, it was this electrode/stimulation methodology that ultimately
led to the first reported case of "useful" sight in blind humans via a
64-electrode (platinum encased in a teflon substrate) intracranial MEA system developed by William Dobelle *et al.* [7]. After the implant, Dobelle's
subjects were able to read "Braille" images perceived as phosphenes that
were reproducibly controlled according to the combination of electrodes
stimulated. Although this approach increased the speed with which Braille
could be read, it was seen only as a transitional indicator that this technology could provide "useful" sight in the blind and was not adopted for
common use.

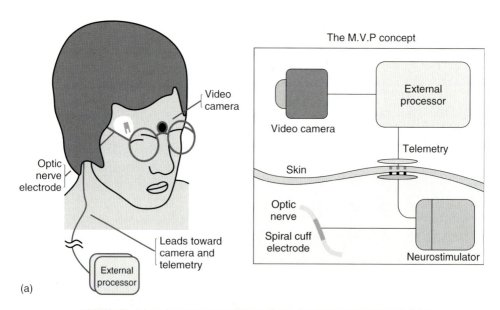

(a)

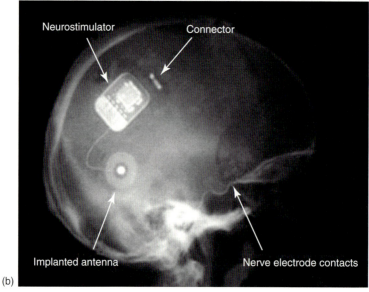

(b)

Figure 1.8 (a) Schematic of visual prosthesis for optic nerve stimulation using a four-channel self-adjusting cuff electrode and (b) X-ray image of the system in place in a blind human. (Adapted from Veraart and coworkers [54].)

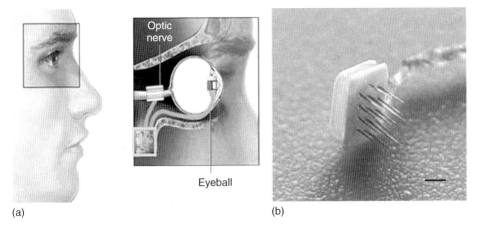

(a)   (b)

Figure 1.9 Schematic of visual prosthesis for optic nerve stimulation using microelectrode arrays from the C-Sight program (a), with 16 probes of 0.2 mm in length (b) scale bar is 1 mm. (Adapted from Boockvar and coworkers [21].)

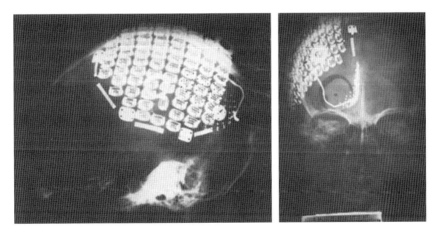

Figure 1.10 The 81-electrode array implanted by Brindley and Lewin. (Adapted from Brindley and Levin [41].)

Dobelle's cortical MEA was also implanted into two humans, one who was able to "see" sufficiently to navigate effectively by sight [55] and one who never perceived phosphenes (Figure 1.11). Nevertheless, while both these people sustained their implants for 20 years without issue, the functioning of the former individual's electrode array eventually deteriorated, requiring removal. This need for a device to be accepted by the surrounding tissues perhaps highlights one of the key challenges facing electrode implants in the human body.

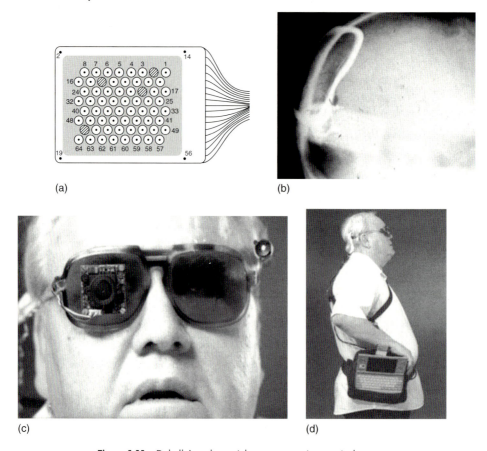

(a) (b)

(c) (d)

Figure 1.11 Dobelle's sub-cranial, nonpenetrating cortical implant (a,b) was the first prosthetic device that restored lost "useful" vision in a human subject (c,d). (Adapted from Dobelle [55].)

Utah Artificial Vision Systems will be designed to essentially transform light from within an incident field of vision captured by a camera and integrate the light, via a pair of spectacles, into electrical signals. These electrical signals will then be translated by a signal processor into electrical stimulation patterns for the visual cortex, that will, in turn, elicit ordered phosphene percepts (25 × 25, or 625 points of stimulation) [56] sufficient to give the sensation of "seeing" objects in the field of "sight." The application of the Utah Electrode Array (UEA) (Figure 1.12) within this system will be to deliver the electrical stimulation to the visual cortex and the current 100 probe configuration will be further developed into a 625 microelectrode configuration [57].

2 mm ━━━━━━   2 mm ━━━━━━

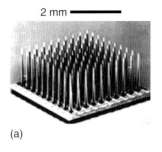 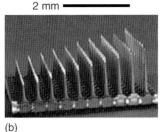 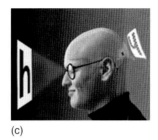

(a)   (b)   (c)

Figure 1.12 (a) The Utah Electrode Array (UEA). (b) The Utah Slanted Electrode Array (USEA). Each of the electrodes in these arrays is made of insulated (silicon) gold wires tipped with platinum (initially) or iridium oxide. They can be implanted in regions of the central or peripheral nervous systems. (c) Schema for cortical stimulation using the UEA in the Utah Artificial Vision System. (Adapted from Normann [57].)

The UEA and Utah Slanted Electrode Array (USEA) consist of gold wires insulated with silicon, tipped with platinum, or, more recently, iridium oxide. Each of the silicon probes are 80 μm in diameter at their base, around 1.5 mm in length (UEA) and spaced from each other at distances of 0.4 mm. The slanted version of the array differs from the UEA by a progressive gradation in the length of the individual probes from 0.5 to 1.5 mm in length (Figure 1.12), primarily for use in peripheral nerves. Essentially, the UEA systems have been designed to integrate with neural tissue, which raises the important question (relevant to any other similar implantable electrode array) as to how they can be removed if they show signs of failure. In particular, how the removal of such electrodes will affect the function of that tissue into which they had been implanted is a major consideration that still needs to be addressed within the medical bionics area.

Within the backdrop of enormous pressure driving rapid technological improvement, significant advances have been made in all aspects of artificial vision technologies. It is at the point of transaction between the neural and electronic circuitry, for example, the electrode/neural tissue interface, that perhaps the most rapid development has taken place since the 1950s. There are proposals to expand Dobelle's original 68-electrode MEA system to 516 electrodes [55] and, as mentioned, the UEA system to at least 625 microelectrodes (the minimum electrode number sufficient to allow reading and effective navigation) [58]. The Michigan Electrode Array has already achieved as many as 1024 microelectrodes [59, 60] but is, at present, limited by the spatial requirements of the tissue into which it is intended to be implanted and availability of sufficient wiring to address each electrode. It is thus not surprising that neuroelectrode fabrication technologies and protocols developed in the highly dynamic area of vision restoration have provided a technological platform for the development of many other neuroprosthetic applications.

1.1.1.3 Neural Prosthetic Applications

MEA systems designed to interact with excitable cell systems will inevitably undergo further development to incorporate significantly greater numbers of individual electrodes by which to stimulate discrete cellular entities. These designs will undoubtedly enhance the resolution of neurostimulation, and/or monitoring signals, and will better meet the challenges of the environment into which they are to be implanted. In view of the major advancements in the spatial resolution with which electrodes can interface with the human nervous system, concurrent technologies in connecting and acquiring the neural signals with the electronic circuitry will also need to improve [61].

With a few exceptions (e.g., DBS in Parkinson's disease, vagal nerve stimulation (VNS)), in most applications of CNS stimulation and monitoring, single electrodes generally provide limited therapeutic opportunities. There are significant advantages afforded by high-density/resolution interfacing between the electrodes and neurons. This is particularly evident for human vision studies that have been a driving force behind the design of highly engineered MEAs, which are much smaller, precise, and effective in their ability to address the tissue, hence improving their functionality.

The elegance of current cortical stimulation MEA designs is exemplified by the UEA [62] and USEA [63] described above. The UEA design accommodates applications beyond the restoration of vision, such as connection of volitional thought toward movement with electronic prosthetic limbs [64]. The first step where the connection of thought with electronic prostheses was successfully achieved was in humans with tetraplegic spinal cord injuries [65]. Using the Neuroport® Array (Cyberkinetics Neurotechnology Systems Inc., Foxborough, MA), a commercial version of the UEA was implanted in patients who then demonstrated their ability to move cursors on a computer screen. This initially promising result has been extended in a recent initiative supported by the Defense Advanced Research Projects Agency (DARPA) to connect volitional movement thoughts with electronic prosthetic limbs that may even be able to deliver touch sensation to people who have had one or more limbs amputated.

The versatile "modular" thin-film three-dimensional array (Michigan 3D Electrode Array) consisting of individual modules containing 16 probes with 8 electrodes per probe (total of 128 electrodes per module) that can be attached to each other to form 3D arrays, containing as many as 1028 electrodes [60] (Figure 1.13), could likewise be applied for prosthetic limb control.

This "many birds with few stones" approach and capability, demonstrated initially by the UEA but nevertheless applicable to most of the existing

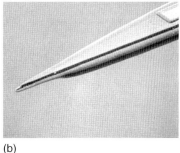

(a)   (b)

Figure 1.13 The Michigan 3D electrode array. A 1028-electrode, 12-channel array (a) and individual probe tip (b). (Adapted from Wise and coworkers [59].)

electrode systems, delivers major advantages in hugely expanding the scope of medical bionics applications able to be effectively targeted by a relatively few devices. This broad applicability to medical bionic devices thus significantly reduces the research and developmental costs required to translate a device into the clinical setting.

1.1.1.4 Vagus Nerve Stimulation (Epilepsy and Pain Management)

Since the earliest observations that stimulation of the vagus nerve (also known as *the pneumogastric nerve, cranial nerve X, the Wanderer,* or *the Rambler*) evoked responses in the ventroposterior complex and the thalamus [66], there has been increasing clinical scientific interest in using this approach for controlling epileptic seizures. This was ultimately confirmed in experiments by Zabara who was able to abort seizures in a strychnine dog model of epilepsy in 1985 [67]. Three years later, five patients with programmable "NeuroCybernetics Prosthesis" stimulators produced by Cyberonics were implanted in humans as part of two pilot studies [68]. These studies showed that electrical stimulation of the vagal nerve led to a 46% mean reduction in the seizures experienced by the tested subjects. After a decade of animal experimentation and clinical trials, in 1997 the vagal nerve stimulator received regulatory approval from the US FDA for use in people older than 12 years of age who are affected by refractory partial-onset seizures. All VNS systems are similar in design. For example, the Cyberonics "NeuroCybernetics Prosthesis" (Figure 1.14) consists of a titanium-encased generator powered by a lithium battery that is implanted within the chest cavity [69]. A lead wire system with platinum–iridium electrodes in a triple coil cuff configuration is wound around the vagus nerve, where it is sutured in place (Figure 1.14c). To date, more than 30 000 people have been treated with VNS [70].

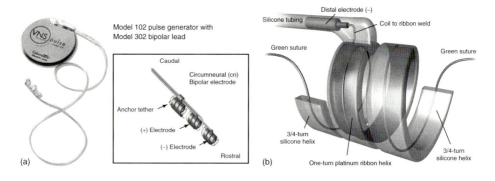

Model 102 pulse generator with Model 302 bipolar lead

Caudal

Circumneural (cn) Bipolar electrode

Anchor tether

(+) Electrode

(−) Electrode

Rostral

(a)

Distal electrode (−)

Silicone tubing

Coil to ribbon weld

Green suture

Green suture

3/4-turn silicone helix

3/4-turn silicone helix

One-turn platinum ribbon helix

(b)

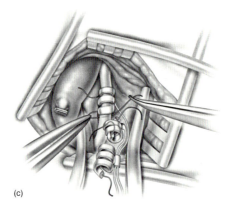

(c)

Figure 1.14 NeuroCybernetics vagal nerve stimulator system with a 102 pulse generator and 302 bipolar cuff electrode (a). (b) Diagram of the distal electrode, and (c) its implantation around the vagus nerve. (Adapted from Santos and [69] and Nemeroff and coworkers [70].)

During early VNS trials, observations were made that the moods of epileptic test subjects improved with stimulation. This led to a potential role for VNS therapy in depression. A focused study of depression symptom severity rating scales in subjects with epilepsy confirmed that VNS therapy was directly associated with positive mood changes [71, 72]. In addition, there was physiological evidence of reductions in the metabolic activity in CNS regions (i.e., amygdala, hippocampus, cingulate gyrus) associated with mood [73]. However, while science indicates benefits in using VNS to treat unresponsive depression, it is generally accepted within the field that there is still much to be learned before VNS is used to treat a depressive disorder. VNS has also been reported to reduce experimentally induced pain [74], and migraine [75], although again, through largely undefined pathways.

1.1.1.5 Transcutaneous Electrical Nerve Stimulation

Devices used for pain management utilize electrical impulses aimed at blocking nociception (pain). For example, surgical placement of epidural electrodes along the dorsal columns allows spinal cord stimulation. This approach is hypothesized to work through the gate-control theory of Wall and Melzack [76], whereby the strategically placed electrodes stimulate the dorsal columns to inhibit the incoming nociceptive input via the so-called A delta or C fibers in the spinal cord.

1.1.1.6 Cardiovascular Applications

Cardiac pacing involves the implantation of a medical device called *a pacemaker* (or *artificial pacemaker*, so as not to be confused with the heart's natural pacemaker), which uses electrical impulses delivered by electrodes to promote timed contraction of heart muscles and thus regulate the beating of the heart (Figure 1.15).

A cardiac pacing lead is just a simple cable that connects the pulse generator to the heart. The size, longevity, and features of the pulse generator and the patient's safety depend on the performance of the lead. The leads are generally thin, insulated wires that are designed to carry electricity between the battery and heart. Depending on the type of pacemaker, it will contain either a single lead for single-chamber pacemakers, or two

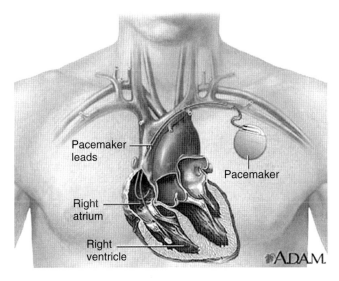

Figure 1.15 Image shows the location of the leads and electronics (including power source) for the implantable artificial pacemaker. (Image obtained with permission from *www.education.science-thi.org.*)

leads for dual-chamber pacemakers. With the constant beating of the heart, these wires are chronically flexed and hence must be resistant to fracture. There are many styles of leads available, with primary design differences found at the exposed end. Many of the leads have a screw-in tip, which helps anchor them to the inner wall of the heart. Pacing leads typically have an electrode pair at the tip comprising a tip electrode and a ring (or band) electrode, both of which are used for sensing intracranial electrical activity and administering low-voltage pacing stimuli (Figure 1.16). These pacing electrodes are commonly made of platinum–iridium (Pt alloyed with 10–20% Ir) [77].

The pacemaker must assure that the heart will contract with each stimulus. Thus, the pacemaker must deliver a controlled pulse of voltage with sufficient duration to assist the patient to cope with any minute-to-minute variations that may occur in their heart beat during daily life. At the same time, it is necessary for the device to use minimal current to maintain battery longevity. The most common approaches to reliably deliver stimuli, within adequate safety parameters while decreasing the current drained from the battery, have traditionally centered around electrode size and material, surface structure and shape and, more recently, glucocorticosteroid sustained release [78].

1.1.1.7 Orthopedic Applications

The use of electrical bone growth stimulation (EBGS) is an accepted clinical approach to promote bone growth and heal fractures. In the early days of electrically induced bone growth, a range of metal electrode materials were investigated. In 1982, Spadaro [79] employed six different metallic cathode materials in a rabbit medullary canal and applied cathodic currents of 0.02 and 0.2 $\mu A\,mm^{-2}$ for 21 days. This stimulation resulted in quantitative differences in new bone growth. That is, platinum, cobalt-chrome (F-90), and silver led to more bone relative to control implants at the lower geometric current density, while stainless steel (316L) and titanium cathodes were more effective at the higher current. On average, there was a significant increase (46–48%) in new bone formation for the active versus control implants, for either current level. In today's market, the electrode material of choice used for internal stimulation is titanium.

There are three types of EBGS available:

1) Noninvasive EBGSs are externally worn devices (Figure 1.16) that generate a weak electric current within the target site, using either pulsed electromagnetic fields, capacitive coupling, or combined magnetic fields.

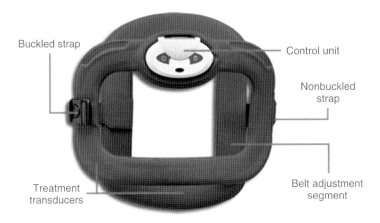

Figure 1.16 Orthofix external bone growth stimulator. (Image obtained with permission from *www.orthofix.com.*)

2) Semi-Invasive (semi-implantable) EBGSs that use direct current supplied by an external power generator and percutaneously placed electrodes.

3) Invasive EBGSs that use direct current require surgical implantation of both the current generator and an electrode (Figure 1.17). Usually, the generator is implanted in an intramuscular or subcutaneous space, and the electrode is implanted within the target bone site. The device typically remains functional for six to nine months after implantation. On completion of treatment, the generator is removed in a second surgical procedure. The electrode may or may not be removed.

OsteoGen®
Bone Growth
Stimulator*

Figure 1.17 OsteoGen internal bone growth stimulator. (Image obtained with permission from *www.biomet.com.*)

One of the most commonly used external stimulating bionic devices on the market today is the transcutaneous electrical nerve stimulation (TENS) system (Figure 1.18a). TENS is a noninvasive, safe nerve stimulation intended to reduce pain, both acute and chronic. While controversy exists as to its effectiveness in the treatment of chronic pain, a number of systematic reviews

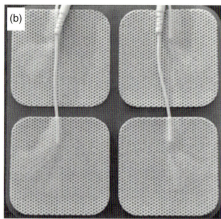

Figure 1.18 (a) A handheld LogiSTIM TN11 TENS machine and (b) associated electrodes. (Image obtained with permission from *www.tens.com.au*.)

or meta-analyses have confirmed its effectiveness for postoperative pain, osteoarthritis, and chronic musculoskeletal pain [80]. The electrode materials used in the TENS system is typically a nonwoven activated carbon cloth (Figure 1.18b). To ensure efficient stimulation to the desired area, a conductive gel is applied to the stimulation area before placement of the electrode.

Another external stimulation device is the electromuscle stimulation (EMS) system, which is often confused with the TENS system. EMS and TENS devices look similar, with both using long electric lead wires and electrodes. However, TENS is used for blocking pain, whereas EMS is used for stimulating muscles. The EMS system is the modern day version of Galvani's electrical myostimulation system in that it applies electrical current to muscles to induce a response. EMS is commonly used for both medical rehabilitation and training purposes, for instance, in the prevention of disuse muscle atrophy, which can occur after musculoskeletal injuries such as damage to bones, joints, muscles, ligaments, and tendons. As such, the EMS provides a philosophical basis for the use of electrical stimulation for promotion of a regenerative response in diseased and/or damaged muscles.

1.2
Key Elements of a Medical Bionic Device

All of the medical bionic devices discussed above comprise stimulating electrodes that are wired to and controlled by electronics within appropriate insulating materials. While advances in each of these components are critical, here we focus on the electrode–cellular interface and the use of organic conductors as alternatives, or complementary materials, to metals. The organic conducting materials used to provide effective cellular communications must, at first, have appropriate electronic properties, provide high or otherwise unfettered conductivity and low impedance in order to minimize the energy drain required to power the system. Low impedance is particularly important for increasing the charge-injection capacity so as to avoid potentially toxic nonfaradaic reactions at the electrode–cellular interface. The materials must also have appropriate mechanical properties. For some applications, flexible electrodes are preferred and so the conductor itself should be flexible or easily fabricated into a flexible device. The chemical properties, including hydrophobicity and hydrophilicity, of the conductor should be biocompatible, noncytotoxic, and sufficiently stable to undergo sterilization without deterioration of the above properties. Long-term material stability (e.g., years) *in vivo* may also be a requisite. For regenerative medical bionics applications, the stability demands are more complex in that the implantable elements of the device should preferably be biodegradable, with no side effects from breakdown products, over an appropriate time frame for the application. Finally and critically, the electrode materials to be used should be processable in such a way that they lend themselves to fabrication into practical devices that can assist the patient.

It is in all of these above aspects of electrode design that organic conducting materials may provide new and exciting opportunities for the development of bionic technologies.

1.2.1
Organic Conductors

As previously discussed, to date, the electrode materials used in bionic device implants include platinum (Pt), iridium (Ir), iridium oxide, and titanium (Ti). Bionic devices need to be able to effectively transmit across the cell–electrode interface [12] and all of these materials have demonstrated this capacity over several decades. However, there remains a need to develop the materials inventory for medical bionics. The prospect of bionic materials that can deliver electrical stimulation, allow integration or attachment of bioactive molecules, and have tunable mechanical properties is an exciting one.

Organic conductors, a class of materials that are conductive and predominantly composed of the element carbon (C), have emerged as excellent candidates for use in medical bionic devices [12]. Carbon can exist in a number of allotropes (different solid forms) commonly known as *diamond, graphite, carbon nanotubes* (single-walled nanotubes, SWNTs and multi-walled nanotubes, MWNTs), or *fullerenes*. The wide-ranging physicochemical properties of these carbons are discussed in more detail in Chapter 2.

Figure 1.19 SJM Masters HP Valved Graft. (Image obtained with permission from *www.sjmprofessional.com*.)

Carbon in the form of pyrolytic carbon (PyC) and diamond-like carbon (DLC) are routinely investigated as implant coatings because of their biocompatibility, superior wear resistance, and chemical inertness. PyC and DLC are often deposited as a coating applied to a substrate. The use of PyC in biomedical applications dates back to the early 1960s when it was used to fabricate heart valves because of its low blood clotting properties. For artificial heart valves that contain both metal and carbon components, vapor-deposited carbon is often used to coat the metallic components and the sewing rings in order to reduce wear and thrombus (blood clot) formation (Figure 1.19). Dacron, polypropylene, and Teflon can also be coated with vapor-deposited carbon for use as vascular grafts. The presence of the thin carbon surface reduces thrombus formation, while maintaining the pliability of the graft [81].

The biocompatibility of DLC was investigated in the early 1990s [82, 83] through cell culture techniques and observing biotolerance after implanting in laboratory animals. Later studies that exposed DLC coatings to human fibroblast and human osteoblast-like cells for several days [84] showed promising results for the use of DLC as a hemocompatible material [85, 86]. Other studies, testing the possibility of DLC as a biocompatible material [87], have indicated that DLC may be an alternative to current materials for biomedical applications. One challenge is that coatings are often subject to severe external forces during an implant's lifetime in the body environment. For example, in medical implants for orthopedic and cardiovascular applications, the body environment can sometimes lead to delamination and spallation of the coatings. To attain superior biocompatible properties adapted for a specific biological function, DLC films have been successfully doped and alloyed with elements such as Si, P, Ti, N, F, Cu [88, 89], and compounds like Ca–O [90].

The introduction of carbon materials for use as implantable electrodes continues to rapidly expand. Most of the research surrounding the use of carbon as the electrode material has focused on lowering the electrode impedance while maintaining biocompatibility. The biocompatibility of carbon for implantable electrodes has been studied using a range of cell types such as nerve [91], muscle (including skeletal [92] and smooth types [93]), and stem cells [94–96].

1.2.1.1 Neural Stimulation and Recording

As has been previously outlined in this chapter, electrical stimulation of nerve cells is widely employed in the development of neural prostheses (including hearing [97], vision [98], mood, and limb movement restoration [99]), with some having translated into clinical therapies (e.g., for Parkinson's disease, dystonia, and chronic pain [100]). In all of these applications, an implanted microelectrode array stimulates the neurons and modulates their behavior. Research into utilizing carbon as an electrode material has resulted in a large number of publications outlining the usefulness of carbon nanotubes [101–103] and carbon fibers in similar applications.

The current state of the art in treating neurological disorders such as epilepsy or Parkinson's disease involves either drugs or open-loop electrical stimulation of neurological tissue. Carbon fibers are increasingly being used as electrodes for recording brain activity [104] because of their lower electrical impedance compared to standard metal electrodes, which provides the advantage of better signal-to-noise resolution. Another advantage of carbon fiber electrodes is the size in which they can be fabricated. Typically, carbon fiber electrodes have smaller geometric dimensions than metal electrodes, providing the ability to obtain brain activity data with finer spatial resolution. This design has been further improved over the years by introducing new methods of etching the carbon fiber tip [105, 106].

Another group of organic conductors, OCPs, have emerged as promising materials for bionic applications [107–109]. OCPs provide versatility in composition, which has thus far been unavailable from traditional conducting materials. Their unique properties and relatively new "entry" into medical bionics [110] is discussed in greater detail in Chapter 3. The most studied of the OCPs are the polypyrroles, polythiophenes, and polyanilines (structures **1**, **2**, and **3**, respectively, in Chapter 3). The ability to produce the polypyrroles at neutral pH from aqueous media has meant that, to date at least, the use of this OCP backbone has attracted most attention from those interested in the development of new materials for medical bionics.

Using polypyrrole, a significant proportion of the material can be rendered "biological" in nature through the incorporation of appropriate biomaterials such as biological polyelectrolytes or even living cells [111, 112]. Some of the biological polyelectrolytes incorporated into conducting polymers during synthesis are shown in Figure 1.20. In all cases, electronically conductive polymers are produced. An interesting and beneficial consequence of incorporating polyelectrolytes as the dopant is that often gellike (high water content) electronic materials are formed spontaneously on formation of the conducting polymer.

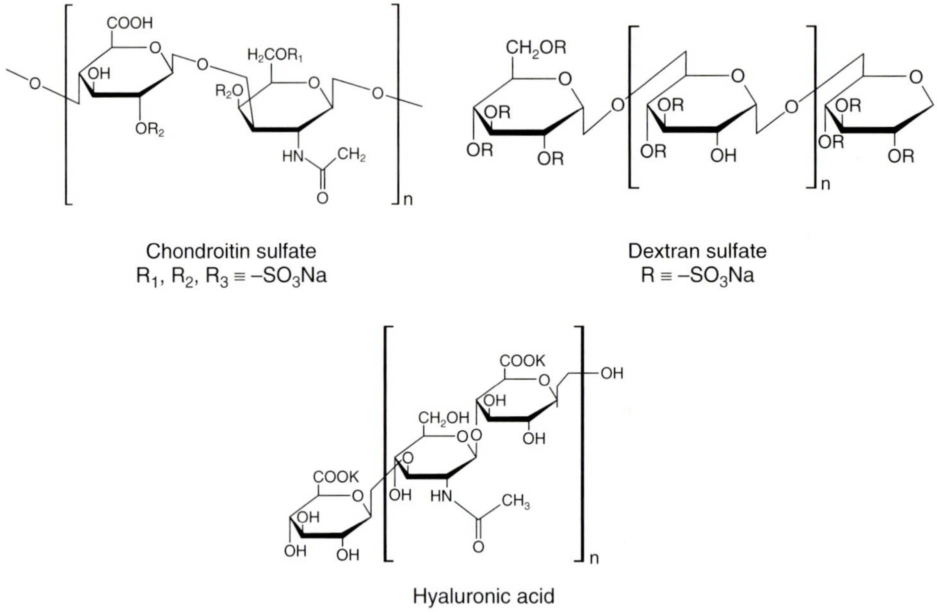

Chondroitin sulfate
$R_1, R_2, R_3 \equiv -SO_3Na$

Dextran sulfate
$R \equiv -SO_3Na$

Hyaluronic acid

Figure 1.20 Biological molecules used as dopants for synthesis of organic conducting polymers. n = number of repeat units and defines the molecular weight.

Another exciting feature of OCPs is the ability to create the electrode assembly under physiological conditions, permitting fabrication in the presence of living cells. Wallace and coworkers [113] employed facile polymerization conditions to incorporate red blood cells into a polypyrrole matrix, while more recently Martin and coworkers [114] developed an ingenious protocol that allows the formation of a conducting polymer in the presence of living cells (neuroblastoma-derived cells) *in vitro*. These novel approaches enable the seamless integration of an electrode with a functional network of living cells. The electronic conductivity of these organic materials, combined with a composition that is biological in nature, provide a unique platform for medical bionic applications. Although a significant step forward, OCPs such as polypyrrole provide further dimensions in our desire to create more effective electrode–cellular interfaces.

1.2.2
Emerging Areas of Application for Medical Bionics

There are a number of treatments for nerve- and muscle-based diseases or injuries, in addition to existing implantable technologies based on nonconductive materials, that could benefit from the introduction of a bionics dimension; some of these are discussed below.

1.2.2.1 Bionics for Peripheral Nerve Injury

It has been estimated that more than 1 million people in the United States suffer serious peripheral nerve injuries arising from lower limb trauma alone every year [115]. On an annual basis, over 50 000 procedures are performed in the United States to correct problems associated with such injuries, costing more than US$7 billion [116]. The incidence of neuropathies or nerve dysfunction due to other pathological processes such as diabetes [117] furthers the socioeconomic impact of peripheral nerve injury.

Thus, a need has emerged for the design of bionic prosthetics that can repair damaged nerve tissue itself, giving rise to the recent design, fabrication, and evaluation of peripheral nerve repair tubes or conduits. Arguably, the most successful peripheral nerve conduits have been fabricated from biological tissue-based materials including nerve, blood vessels, and amniotic tubes [118–121]. Overall, these materials have been incorporated in various combinations or on their own into nerve repair conduits as electrospun webs [121], fibers [122], sponges [123], and hydrogels [124], but again, these structures do not employ OCPs. Likewise, conduits provide an alternative for the treatment of acquired nerve defects such as diabetic neuropathy. Nevertheless, there are still considerable challenges and room for improvement in the development of synthetic conduits for nerve repair, particularly for longer nerve repair applications [125], and selective incorporation of OCPs into this landscape may provide appropriate augmentative alternatives that may facilitate greater control of neuroregenerative response in peripheral nerve injury.

Improvement of nerve guides for effective repair of longer nerve lesions has focused on the incorporation of neurotrophins [126], but for peripheral nerve repair, factor release through electrical stimulation has by and large been overlooked in existing applications under development. Neurotrophins and other bioactive molecules provide further options for the control, extent, and nature of conduit-mediated nerve repair, and OCPs may be the most appropriate underlying materials platform by which complex multimodal delivery paradigms may be developed and employed. In particular, neurotrophic molecular factors such as neurotrophin-3 (NT-3) [127] and brain-derived neurotrophic factor (BDNF) [128] provide neuroprotective effects. While some of these factors have been successfully incorporated into OCPs such as polypyrrole, and subsequently shown to be able to promote nerve growth *in vitro* and *in vivo* [129–131], other such factors require refinement of the fabrication technologies to enable their incorporation into OCPs.

In conjunction with the ability to promote electrically stimulated release, more or less "on demand," these specificities of neurobiological activation provide an opportunity to tailor the control of nerve repair with multicomponential conduits that contain multiple factors that "build" the regenerative environment to a tightly controlled bioreactive and biointeractive chamber. This opportunity to specify axonal regeneration can be further augmented by the development of OCP-mediated trophic control of neuroglial nerve components. Finally, advances in polymer fabrication technologies in recent times have enabled the fabrication of scaffolds with combined therapeutic modalities, as highlighted by coincorporation of glial growth factor (GGF) and Schwann cells within a nerve repair conduit [132], although, again, these studies did not utilize OCPs. It is nevertheless evident that incorporation of OCPs within contemporary peripheral nerve repair conduits stands to vastly improve nerve repair outcomes in the future.

1.2.2.2 Bionics for Damaged or Diseased Muscle

As a tissue composed of excitable cells, muscle bears similarities in the applicability of OCPs to the repair of damaged and diseased muscle tissue. There are a number of target areas that OCPs may add value to the restoration of function in muscle tissues. These include delivery of factors that generate a promyogenic environment in damaged or diseased muscle to enable effective myoregeneration. OCPs may also act as conduits for electrical stimulation to promote muscle regeneration.

As a clinical group, neuromuscular disorders (NMDs) are a heterogeneous set of hereditary disorders that are collectively characterized by progression of clinical symptoms and ongoing muscle degeneration (loss), which in some cases leads to premature death. Indeed, the electrical myostimulation pioneer Duchenne is credited with the characterization of one of the most severe NMDs, Duchenne muscular dystrophy. The common denominator of progressive muscle loss in NMDs can be caused by degenerative effects in muscle directly or by degenerative changes in the nerves innervating the muscle, which result in a lack of physiological challenge to and the subsequent breakdown of the muscle.

Myoregenerative factors such as leukemia inhibitory factor (LIF), and insulin-like growth factor 1 (IGF-1) and the myoregulatory growth differentiating factor 8 (GDF-8; i.e., myostatin) are known to improve cell-mediated muscle remodeling in dystrophic muscles [133–136]. Furthermore, cell-mediated muscle remodeling is vastly improved in the presence of controlled release of one or another of these factors [133] compared to cells injected alone. From such studies, it is also evident that pretreatment (conditioning) of cells with one or another of these factors before implantation improves the myoregenerative response of the cells when introduced into the degenerating muscle. These findings have been recently highlighted by improvements in transplanted cell engraftment by coinjection of nerve growth factor (NGF) in the dystrophic muscle of *mdx* mice [137].

Observed beneficial promyogenic effects of electrical stimulation [138] and mechanical stress on muscle, and subsequently on myogenic development, involve enhanced myoregeneration [139] and activation of promyogenic pathways [140, 141], including upregulation of growth factors such as IGF-1 [141]. The future development of polymer-based electrode systems to assist in the regeneration of muscle heralds an exciting new dawn of organic bionics that will undoubtedly benefit from advances in electromaterials in the near, mid, and long term.

1.2.3
Outline of the Book

Here, we introduce the reader to the world of organic conductors and their potential application in the development of highly effective medical bionic devices.

In particular, we focus on the emerging field of nanostructured carbons in Chapter 2 and OCPs in Chapter 3. Biological studies involving these materials are discussed in Chapter 4.

As with all areas of materials science development, the field is confronted by the issues of processing and device fabrication. We tackle the challenges and opportunities in front of us at this point in time in Chapter 5.

Finally in Chapter 6 – with eyes wide open – we dream of what might be in the not too distant future.

References

1. Clark, G.M. (2003) *Sounds from Silence: Graeme Clark and the Bionic Ear Story*, Allen & Unwin, New York.
2. Roth, B.J. (1994) Mechanisms for electrical stimulation of excitable tissue. *Crit. Rev. Biomed. Eng.*, **22**, 253–305.
3. Kolff, W.J., Akutsu, T., Dreyer, B., and Norton, S.H. (1959) Artificial heart in the chest and use of polyurethane of making hearts, valves and aortas. *Trans. Am. Soc. Artif. Intern. Organs*, **7**, 298–300.
4. Kolff, W.J. (1969) The artificial heart: research, development or invention? *Dis. Chest*, **56**, 314–329.
5. Zoll, P.M., Linenthal, A.J., Gibson, W., Paul, M.H., and Norman, L.R. (1956) Termination of ventricular fibrillation in man by externally applied electric countershock. *N. Engl. J. Med.*, **254**, 727–732.
6. Elmquist, R. and Senning, A. (1960) An implantable pacemaker for the heart, in *Medical Electronics: Proceedings of the 2nd International Conference on Medical Electronics* (ed. C.N. Smyth), Iliffe & Sons, London, p. 253.
7. Dobelle, W.H., Mladejovsky, M.G., Evans, J.R., Roberts, T.S., and Girvin, J.P. (1976) Braille reading by a blind volunteer by visual cortex stimulation. *Nature*, **259**, 111–112.
8. Clark, G.M., Tong, Y.C., Black, R., Forster, I.C., Patrick, J.F., and Dewhurst, D.J. (1977) A multiple electrode cochlear implant. *J. Laryngol. Otol.*, **91** (11), 935–945.
9. Clark, G.M. (1978) Cochlear implant surgery for profound or total

hearing loss. *Med. J. Aust.*, **2** (13), 587–588.

10. Benabid, A.L., Pollak, P., Louveau, A., Henry, S., and de Rougemont, J. (1987) Combined (thalamotomy and stimulation) stereotactic surgery of the VIM thalamic nucleus for bilateral Parkinson disease. *Appl. Neurophysiol.*, **50**, 344–346.

11. Kuiken, T.A., Dumanian, G.A., Lipschutz, R.D., Miller, L.A., and Stubblefield, K.A. (2004) The use of targeted muscle reinnervation for improved myoelectric prosthesis control in a bilateral shoulder disarticulation amputee. *Prosthet. Orthot. Int.*, **28**, 245–253.

12. Wallace, G.G., Moulton, S.E., and Clark, G.M. (2009) Electrode-cellular Interface. *Science*, **324**, 185–186.

13. Hubel, D.H. (1957) Tungsten microelectrode for recording from single units. *Science*, **125**, 549–550.

14. Campbell, P.K., Jones, K.E., and Normann, R.A. (1990) A 100 electrode intracortical array: structural variability. *Biomed. Sci. Instrum.*, **26**, 161–165.

15. Campbell, P.K., Jones, K.E., Huber, R.J., Horch, K.W., and Normann, R.A. (1991) A silicon-based, three-dimensional neural interface: manufacturing processes for an intracortical electrode array. *IEEE Trans. Biomed. Eng.*, **38**, 758–768.

16. Pellinen, D., Moon, T., Vetter, R., Miriani, R., and Kipke, D. (2005) Multifunctional flexible parylene-based intracortical microelectrodes. *Conf. Proc. IEEE Eng. Med. Biol. Soc.*, **5**, 5272–5275.

17. Jensen, W., Yoshida, K., and Hofmann, U.G. (2006) In-vivo implant mechanics of flexible, silicon-based ACREO microelectrode arrays in rat cerebral cortex. *IEEE Trans. Biomed. Eng.*, **53**, 934–940.

18. Loeb, G.E., Peck, R.A., and Martyniuk, J. (1995) Toward the ultimate metal microelectrode. *J. Neurosci. Methods*, **63**, 175–183.

19. Keefer, E.W., Botterman, B.R., Romero, M.I., Rossi, A.F., and Gross, G.W. (2008) Carbon nanotube coating improves neuronal recordings. *Nat. Nanotechnol.*, **3**, 434–439.

20. Cogan, S.F. (2008) Neural stimulation and recording electrodes. *Annu. Rev. Biomed. Eng.*, **10**, 275–309.

21. Boockvar, J.A., Telfeian, A., Baltuch, G.H., Skolnick, B., Simuni, T., Stern, M., Schmidt, M.L., and Trojanowski, J.Q. (2000) Long-term deep brain stimulation in a patient with essential tremor: clinical response and postmortem correlation with stimulator termination sites in ventral thalamus. Case report. *J. Neurosurg.*, **93**, 140–144.

22. Wise, K.D. (2005) Silicon microsystems for neuroscience and neural prostheses: interfacing with the central nervous system at the cellular level. *IEEE Eng. Med. Bio. Mag.*, **24** (5), 22–29.

23. Sui, X., Li, L., Chai, X., Wu, K., Zhou, C., Sun, X., Xu, X., Li, X., and Qiushi, R. (2009) Visual prosthesis for optic nerve stimulation, in *Implantable Neural Prostheses 1* (eds D.D. Zhou and E. Greenbaum), Springer, pp. 43–85.

24. Epstein, J. (1989) *The Story of the Bionic Ear*, Hyland House, Australia.

25. Gisselsson, L. (1950) Experimental investigation into the problem of humoral transmission in the cochlea. *Acta Otolaryngol. Suppl.*, **82**, 9–78.

26. Djourno, A. and Eyries, C. (1957) Auditory prosthesis by means of a distant electrical stimulation of the sensory nerve with the use of an indwelt coiling. *Presse Med.*, **35**, 14–17.

27. Merzenich, M.M., Michelson, R.P., Petit, C.R., Schindler, R.A., and Reid, M. (1973) Neural recording of sound sensation evoked by electrical stimulation of the acoustic nerve. *Ann. Otol. Rhinol. Laryngol. Suppl.*, **82**, 486–503.

28. House, W.F. and Urban, J. (1973) Long term results of electrode implantation and electronic stimulation of the cochlea in man. *Ann. Otol. Rhinol. Laryngol.*, **82**, 504–517.

29. House, W.F. (1976) Cochlear implants. *Ann. Otol. Rhinol. Laryngol. Suppl.*, **85** (Suppl. 27), 1–93.

30. Graeme, Clark, Cowan, R.S.C. and Dowell, R.C. (1997) *Cochlear Implantation for Infants and Children: Advances*, Singular Press, San Diego.

31. Le Roy, C. (1755) Ou l'on rend compte de quelques tentatives que lon a faites pour guerir plusieurs maladies par l'ectricite. *Hist. Acad. R. Sci. (Paris), Mem. Math. Phys.*, **60**, 87–95.

32. Marg, E. (1991) Magnetostimulation of vision: direct noninvasive stimulation of the retina and the visual brain. *Optom. Vis. Sci.*, **68** (6), 427–440.

33. Pascual-Leone, A. and Wagner, T. (2007) A brief summary of the history of noninvasive brain stimulation. *Annu. Rev. Biomed. Eng.*, **9**, 527–565.

34. Rizzo, J.F. and Wyatt, J. (1997) Prospects of visual prosthesis. *Neuroscientist*, **3**, 251–262.

35. Liu, W., McGucken, E., Vichienchom, K., Clements, S.M., Demarco, S.C., Humayun, M., de Juan, E., Weiland, J., and Greenberg, R. (1999) Retinal prosthesis to aid the visually impaired. *Proc. IEEE Int. Conf. Syst. Man Cybernet.*, **4**, 364–369.

36. Humayun, M.S., Weiland, J.D., Fujii, G.Y., Greenberg, R., Williamson, R., Little, J., Mech, B., Cimmarusti, V., Van Boemel, G., Dagnelie, G., and de Juan, E.J. (2003) Visual perception in a blind subject with a chronic microelectronic retinal prosthesis. *Vision Res.*, **43**, 2573–2581.

37. Veraart, C., Raftopoulos, C., Mortimer, J.T., Delbeke, J., Pins, D., Michaux, G., Vanlierde, A., Parrini, S., and Wanet-Defalque,

M.C. (1998) Visual sensations produced by optic nerve stimulation using an implanted self-sizing spiral cuff electrode. *Brain Res.*, **813**, 181–186.

38. Delbeke, J., Oozeer, M., and Claude, V. (2003) Position, size and luminosity of phosphenes generated by direct optic nerve stimulation. *Vision Res.*, **43**, 1091–1102.

39. Dobelle, W.H., Mladejovsky, M.G., and Girvin, J.P. (1974) Artificial vision for the blind: electrical stimulation of visual cortex offers hope for a functional prosthesis. *Science*, **183**, 440–444.

40. Normann, R.A., Maynard, E.M., Rousche, P.J., and Warren, D.J. (1999) A neural interface for a cortical vision prosthesis. *Vision Res.*, **39**, 2577–2587.

41. Brindley, G.S. and Lewin, W.S. (1968) The sensations produced by electrical stimulation of the visual cortex. *J. Physiol. (Lond.)*, **196**, 479–493.

42. Kanda, H., Morimoto, T., Fujikado, T., Tano, Y., Fukuda, Y., and Sawai, H. (2004) Electrophysiological studies of the feasibility of suprachoroidal-transretinal stimulation for artificial vision in normal and RCS rats. *Invest. Ophthalmol. Vis. Sci.*, **45**, 560–566.

43. Humayun, M., Propst, R., de Juan, E. Jr., McCormick, K., and Hickingbotham, D. (1994) Bipolar surface electrical stimulation of the vertebrate retina. *Arch. Ophthalmol.*, **112** (1), 110–116.

44. Grumet, A.E., Wyatt, J.L., and Rizzo, J.F. (2000) Multi-electrode stimulation and recording in the isolated retina. *J. Neurosci. Methods*, **101** (1), 31–42.

45. Suzuki, S., Humayun, M.S., Weiland, J.D., Chen, S.J., Margalit, E., Piyathaisere, D.V., and de Juan, E. (2004) Comparison of electrical stimulation thresholds in normal and retinal degenerated mouse retina. *Jpn. J. Ophthalmol.*, **48** (4), 345–349.

46. Humayun, M.S., de Juan, E. Jr., Dagnelie, G., Greenberg, R.J., Propst, R.H., and Phillips, D.H. (1996) Visual perception elicited by electrical stimulation of retina in blind humans. *Arch. Ophthalmol.*, **114** (1), 40–46.

47. Humayun, M.S., de Juan, E.J., Weiland, J.D., Dagnelie, G., Katona, S., Greenberg, R., and Suzuki, S. (1999) Pattern electrical stimulation of the human retina. *Vision Res.*, **39**, 2569–2576.

48. Rizzo, J.F. III, Wyatt, J., Loewenstein, J., Kelly, S., Shire, D., and Wyatt, J. (2003) Methods and perceptual thresholds for short-term electrical stimulation of human retina with microelectrode arrays. *Invest. Ophthalmol. Vis. Sci.*, **44** (12), 5355–5361.

49. Rizzo, J.F. III, Wyatt, J., Lowenstein, J., Kelly, S., and Shire, D. (2003) Perceptual efficacy of electrical stimulation of human retina with a microelectrode array during short term surgical trials. *Invest. Ophthalmol. Vis. Sci.*, **44** (12), 5362–5369.

50. Weiland, J.D., Liu, Y., and Humayun, M.S. (2005) Retinal prosthesis. *Annu. Rev. Biomed. Eng.*, **7**, 361–401.

51. Chader, G.J., Weiland, J., and Humayun, M.S. (2009) Artificial vision: needs, functioning, and testing of a retinal electronic prosthesis. *Prog. Brain Res.*, **175**, 317–332.

52. Humayun, M.S., Weilanda, J.D., Fujiia, G.Y., Greenberg, R., Williamson, R., Little, J., Mech, B., Cimmarustib, V., Van Boemela, G., Dagneliec, G., and de Juan, E. Jr. (2003) Visual perception in a blind subject with a chronic microelectronic retinal prosthesis. *Vision Res.*, **43** (24), 2573–2581.

53. Veraart, C., Raftopoulos, C., Mortimer, J.T. *et al.* (1998) Visual sensations produced by optic nerve stimulation using an implanted self-sizing spiral cuff electrode. *Brain Res.*, **813**, 181–186.

54. Veraart, C., Wanet-Defalque, M.-C., Gérard, B., Vanlierde, A., and Delbeke, J. (2003) Pattern recognition with the optic nerve visual prosthesis. *Artif. Organs*, **27** (11), 996–1004.

55. Dobelle, W.H. (2000) Artificial vision for the blind by connecting a television camera to the visual cortex. *ASAIO J.*, **46**, 3–9.

56. Maynard, E.M. (2001) Visual prosthesis. *Annu. Rev. Biomed. Eng.*, **3**, 145–168.

57. Normann, R.A. (2007) Technology insight: future neuroprosthetic therapies for disorders of the nervous system. *Nat. Clin. Pract. Neuro.*, **3** (8), 444–452.

58. Cha, K., Horch, K.W., Normann, R.A., and Boman, D.K. (1992) Reading speed with a pixelized vision system. *J. Opt. Soc. Am. A*, **9**, 673–677.

59. Wise, K.D., Anderson, D.J., Hetke, J.F., Kipke, D.R., and Najafi, K. (2004) Wireless implantable microsystems: high-density electronic interfaces to the nervous system. *IEEE Proc.*, **92** (1), 76–97.

60. Wise, K.D. (2005) Silicon microsystems for neuroscience and neural prostheses: interfacing with the central nervous system at the cellular level. *IEEE Eng. Med. Bio. Mag.*, **24** (5), 22–29.

61. Wyatt, J.L., Standley, D.L., and Yang, W. (1991) The MIT vision chip project: analog VLSI systems for fast image acquisition and early vision processing. *IEEE Proc.*, **2**, 1330–1335.

62. Jones, K.E., Campbell, P.K., and Normann, R.A. (1992) A glass/silicon composite intracortical electrode array. *Ann. Biomed. Eng.*, **20**, 423–437.

63. Branner, A. and Normann, R.A. (2000) A multielectrode array for intrafascicular recording and stimulation in sciatic nerve of cats. *Brain Res. Bull.*, **51**, 293–306.

64. Normann, R.A. (2007) Technology insight: future neuroprosthetic therapies for disorders of the

nervous system. *Nat. Clin. Pract. Neuro.*, **3**, 444–452.

65. Hochberg, L.R., Serruya, M.D., Friehs, G.M., Mukand, J.A., Saleh, M., Caplan, A.H., Branner, A., Chen, D., Penn, R.D., and Donoghue, J.P. (2006) Neuronal ensemble control of prosthetic devices by a human with tetraplegia. *Nature*, **442** (7099), 164–171.

66. Bailey, P. and Bremer, F.A. (1938) Sensory cortical representation of the vagus nerve. *J. Neurophysiol.*, **1**, 405–412.

67. Zabara, J. (1985) Peripheral control of hypersynchronous discharge in epilepsy. *Electroencephalography*, **61**, S162.

68. Uthman, B.M., Wilder, B.J., Hammond, E.J., and Reid, S.A. (1990) Efficacy and safety of vagus nerve stimulation in patients with complex partial seizures. *Epilepsia*, **31** (Suppl. 2), 44–50.

69. Santos, M. (2004) Surgical placement of the vagus nerve stimulator. *Oper. Tech. Ototlaryngol.*, **15**, 201–209.

70. Nemeroff, C.B., Mayberg, H.S., Krahl, S.E., McNamara, J., Frazer, A., Henry, T.R., George, M.S., Charney, D.S., and Brannan, S.K. (2006) VNS therapy in treatment-resistant depression: clinical evidence and putative neurobiological mechanisms. *Neuropsychopharmacology*, **31**, 1345–1355.

71. Elger, G., Hoppe, C., Falkai, P., Rush, A.J., and Elger, C.E. (2000) Vagus nerve stimulation is associated with mood improvements in epilepsy patients. *Epilepsy Res.*, **42**, 203–210.

72. Harden, C.L., Pulver, M.C., Ravdin, L.D., Nikolov, B., Halper, J.P., and Labar, D.R. (2000) A pilot study of mood in Epilepsy patients treated with vagus nerve stimulation. *Epilepsy Behav.*, **1**, 93–99.

73. Henry, T.R., Bakay, R.A.E., Votaw, J.R., Pennell, P.B., Epstein, C.M., Faber, T.L., Grafton, S.T., and Hoffman, J.M. (1998) Brain blood flow alterations induced by therapeutic vagus nerve stimulation in partial epilepsy: I. Acute effects at high and low levels of stimulation. *Epilepsia*, **39**, 983–990.

74. Kirchner, A., Birklein, F., Stefan, H., and Handwerker, H.O. (2005) Left vagus nerve stimulation suppresses experimentally induced pain. *Neurology*, **55**, 1167–1171.

75. Mauskop, A. (2005) Vagus nerve stimulation relieves chronic refractory migraine and cluster headaches. *Cephalalgia*, **25** (2), 82–86.

76. Wall, P.D. and Melzack, R. (1965) Pain mechanisms: a new theory. *Science*, **150**, 971–979.

77. DiFilippo, F.P. and Brunken, R.C. (2005) Do implanted pacemaker leads and ICD leads cause metal-related artifact in cardiac PET/CT? *J. Nucl. Med.*, **46** (3), 436–443.

78. Stokes, K. (2002) Cardiac pacing electrodes. *IEEE Proc.*, **84** (3), 457–467.

79. Spadaro, J.A. (1982) Electrically enhanced osteogenesis at various metal cathodes. *J. Biomed. Mater. Res.*, **16**, 861–873.

80. Johnson, M. and Martinson, J.M. (2007) Efficacy of electrical nerve stimulation for chronic musculoskeletal pain: a meta-analysis of randomized controlled trials. *Pain*, **130** (1), 157–165.

81. More, R.B., Sines, S., Ma, L., and Bokros, J.C. (2008) Pyrolitic carbon, in *Encyclopedia of Biomaterials and Biomedical Engineering* (eds G.L. Bowlin and G. Wnek), Taylor & Francis Group, London, pp. 2370–2381.

82. Thomson, L.A., Law, F.C., Rushton, N., and Franks, J. (1991) Biocompatibility of diamond-like carbon coating. *Biomaterials*, **12**, 37–40.

83. Mitura, E., Mitura, S., Niedzielski, P., Has, Z., Wolowiec, R., Jakubowski, A., Szmidt, J., Sokolowska, A., Louda, P., Marciniak, J., and Koczy, B. (1994) Diamond-like carbon coatings for

biomedical applications. *Diamond Relat. Mater.*, **3**, 896–898.

84. Butter, R., Allen, M., Chandra, L., Lettington, A.H., and Rushton, N. (1995) In vitro studies of DLC coatings with silicon intermediate layer. *Diamond Relat. Mater.*, **4**, 857–861.

85. Yu, L.J., Wang, X., Wang, X.H., and Liu, X.H. (2000) Haemocompatibility of tetrahedral amorphous carbon films. *Surf. Coat. Technol.*, **128/129**, 484–488.

86. Jones, M.I., McColl, I.R., Grant, D.M., Parker, K.G., and Parker, T.L. (1999) Haemocompatibility of DLC and TiC – TiN interlayers on titanium. *Diamond Relat. Mater.*, **8**, 457–462.

87. Sheeja, D., Tay, B.K., and Nung, L.N. (2004) Feasibility of diamond-like carbon coatings for orthopaedic applications. *Diamond Relat. Mater.*, **13**, 184–190.

88. Maguire, P.D., McLaughlin, J.A., Okpalugo, T.I.T., Lemoine, P., Papakonstantinou, P., McAdams, E.T., Needham, M., Ogwu, A.A., Ball, M., and Abbas, G.A. (2005) Mechanical stability, corrosion performance and bioresponse of amorphous diamond-like carbon for medical stents and guidewires. *Diamond Relat. Mater.*, **14**, 1277–1288.

89. Hauert, R. (2003) A review of modified DLC coatings for biological applications. *Diamond Relat. Mater.*, **12**, 583–589.

90. Dorner-Reisel, A., Schürer, C., Nischan, C., Seidel, O., and Müller, E. (2002) Diamond-like carbon: alteration of the biological acceptance due to Ca-O incorporation. *Thin Solid Films*, **420–421**, 263–268.

91. Seidlits, S.K., Lee, J.Y., and Schmidt, C.E. (2008) Nanostructured scaffolds for neural applications. *Nanomedicine*, **3**, 183–199.

92. Mohanty, M., Anilkumar, T.V., Mohanan, P.V., Muraleedharan, C.V., Bhuvaneshwar, G.S., Derangere, F., Sampeur, Y., and Suryanarayanan, R. (2002) Long term tissue response to titanium coated with diamond like carbon. *Biomol. Eng.*, **19**, 125–128.

93. Bacakova, L., Stary, V., Kofronova, O., and Lisa, V. (2001) Polishing and coating carbon fiber-reinforced carbon composites with a carbon-titanium layer enhances adhesion and growth of osteoblast-like MG63 cells and vascular smooth muscle cells *in vitro*. *J. Biomed. Mater. Res.*, **54**, 567–578.

94. Jan, E. and Kotov, N.A. (2007) Successful differentiation of mouse neural stem cells on layer-by-layer assembled single-walled carbon nanotube composite. *Nano Lett.*, **7** (5), 1123–1128.

95. Chao, T.I., Xiang, S., Chen, C.S., Chin, W.C., Nelson, A.J., Wang, C., and Lu, J. (2009) Carbon nanotubes promote neuron differentiation from human embryonic stem cells. *Biochem. Biophys. Res. Commun.*, **384** (4), 426–430.

96. Mooney, E., Dockery, P., Greiser, U., Murphy, M., and Barron, V. (2008) Carbon nanotubes and mesenchymal stem cells: biocompatibility, proliferation and differentiation. *Nano Lett.*, **8** (8), 2137–2143.

97. Clark, G.M. (2006) The multiple-channel cochlear implant: the interface between sound and the central nervous system for hearing, speech, and language in deaf people – a personal perspective. *Philos. Trans. R. Soc. Lond., Ser. A*, **361**, 791–810.

98. Wong, Y.T., Dommel, N., Preston, P., Hallum, L.E., Lehmann, T., Lovell, N.H., and Suaning, G.J. (2007) Retinal neurostimulator for a multifocal vision prosthesis. *IEEE Trans. Neural. Syst. Rehabil. Eng.*, **15**, 425–434.

99. Navarro, X., Krueger, T.B., Lago, N., Micera, S., Stieglitz, T., and Dario, P.J. (2005) A critical review of Interfaces with the peripheral nervous system for the control of neuroprostheses and hybrid bionic

systems. *Peripher. Nerv. Syst.*, **10**, 229–258.

100. Benabid, A.L. (2003) Deep brain stimulation for Parkinson's disease. *Curr. Opin. Neurobiol.*, **13**, 696–706.

101. Su, H.C., Lin, C.M., Yen, S.J., Chen, Y.C., Chen, C.H., Yeh, S.R., Fang, W., Chen, H., Yao, D.J., Chang, Y.C., and Yew, T.R. (2010) A cone-shaped 3D carbon nanotube probe for neural recording. *Biosens. Bioelectron.*, **26** (1), 220–227.

102. Lin, C.-M., Lee, Y.-T., Yeh, S.-R., and Fang, W. (2009) Flexible carbon nanotubes electrode for neural recording. *Biosens. Bioelectron.*, **24** (9), 2791–2797.

103. Harrison, B.S. and Atala, A. (2007) Carbon nanotube applications for tissue engineering. *Biomaterials*, **28** (2), 344–353.

104. Inagaki, K., Heiney, S.A., and Blazquez, P.M. (2009) Method for the construction and use of carbon fiber multibarrel electrodes for deep brain recordings in the alert animal. *J. Neurosci. Methods*, **178**, 255–262.

105. Armstrong-James, M., Fox, K., and Millar, J. (1980) A method for etching the tips of carbon fibre microelectrodes. *J. Neurosci. Methods*, **2**, 431–432.

106. Kuras, A. and Gutmaniene, N. (2000) Technique for producing a carbon-fibre microelectrode with the fine recording tip. *J. Neurosci. Methods*, **96**, 143–146.

107. Wallace, G.G. and Spinks, G. (2007) Conducting Polymers – bridging the bionic interface. *Soft Matter*, **3**, 665–671.

108. Wallace, G.G. and Moulton, S.E. (2009) Organic bionics: molecules, materials and medical devices. *Chem. Aust.*, **76** (5), 3–8.

109. Wallace, G.G. and Spinks, G.M. (2007) Conducting polymers – a bridge across the bionic interface. *Chem. Eng. Prog.*, **103**, S18–S24.

110. Clark, G.M. and Wallace, G.G. (2004) Bionic ears: their development and future advances using neurotrophins and organic conducting polymers. *Appl. Bionics Biomech.*, **1**, 67–89.

111. Wallace, G.G. and Kane-Maguire, L.A.P. (2002) Manipulating and monitoring biomolecular interactions with conductive electroactive polymers. *Adv. Mater.*, **14**, 953–960.

112. Wallace, G.G., Spinks, G.M., Kane-Maguire, L.A.P., and Teasdale, P.R. (2008) *Conductive Electroactive Polymers: Intelligent Polymer Systems*, 3rd edn, CRC Press, Taylor & Francis Group, Boco Raton.

113. Campbell, T.E., Hodgson, A.J., and Wallace, G.G. (1999) Incorporation of erythrocytes into polypyrrole to form the basis of a biosensor to screen for rhesus (D) blood groups and rhesus (D) antibodies. *Electroanalysis*, **11** (4), 215–222.

114. Richardson-Burns, S.M., Hendricks, J.L., Foster, B., Povlich, L.K., Kim, D.H., and Martin, D.C. (2007) Polymerization of the conducting polymer poly(3,4-ethylenedioxythiophene) (PEDOT) around living neural cells. *Biomaterials*, **28**, 1539–1552.

115. Taylor, C.A., Braza, D., Rice, J.B., and Dillingham, T. (2008) The incidence of peripheral nerve injury in extremity trauma. *Am. J. Phys. Med. Rehabil.*, **87**, 381–385.

116. Evans, G.R. (2001) Peripheral nerve injury: a review and approach to tissue engineered constructs. *Anat. Rec.*, **263**, 396–404.

117. Zochodne, D.W., Guo, G.F., Magnowski, B., and Bangash, M. (2007) Diabetic neuropathy: electrophysiological and morphological study of peripheral nerve degeneration and regeneration in transgenic mice that express IFNβ in β cells. *Diabetes Metab. Res. Rev.*, **23**, 490–496.

118. Lloyd, B.M., Luginbuhl, R.D., Brenner, M.J., Rocque, B.G., Tung, T.H., Myckatyn, T.M., Hunter, D.A., Mackinnon, S.E., and Borschel, G.H. (2007) Use of

motor nerve material in peripheral nerve repair with conduits. *Microsurgery*, **27**, 138–145.

119. Keskin, M., Akbaþ, H., Uysal, O.A., Canan, S., Ayyldz, M., Aar, E., and Kaplan, S. (2004) Enhancement of nerve regeneration and orientation across a gap with a nerve graft within a vein conduit graft: a functional, stereological, and electrophysiological study. *Plast. Reconstr. Surg.*, **113**, 1372–1379.

120. Mohammad, J.A., Warnke, P.H., Pan, Y.C., and Shenaq, S. (2000) Increased axonal regeneration through a biodegradable amnionic tube nerve conduit: effect of local delivery and incorporation of nerve growth factor/hyaluronic acid media. *Ann. Plast. Surg.*, **44**, 59–64.

121. Yao, L., O'Brien, N., Windebank, A., and Pandit, A.J. (2009) Orienting neurite growth in electrospun fibrous neural conduits. *Biomed. Mater. Res. B: Appl. Biomater.*, **90B** (2), 483–491.

122. Kim, Y.T., Haftel, V.K., Kumar, S., and Bellamkonda, R.V. (2008) The role of aligned polymer fiber-based constructs in the bridging of long peripheral nerve gaps. *Biomaterials*, **29**, 3117–3127.

123. Toba, T., Nakamura, T., Lynn, A.K., Matsumoto, K., Fukuda, S., Yoshitani, M., Hori, Y., and Shimizu, Y. (2002) Evaluation of peripheral nerve regeneration across an 80-mm gap using a polyglycolic acid (PGA) – collagen nerve conduit filled with laminin-soaked collagen sponge in dogs. *Int. J. Artif. Organs*, **25**, 230–237.

124. Pfister, L.A., Papaloizos, M., Merkle, H.P., and Gander, B. (2007) Hydrogel nerve conduits produced from alginate/chitosan complexes. *J. Biomed. Mater. Res.*, **80**, 932–937.

125. Hudson, T.W., Evans, G.R., and Schmidt, C.E. (2000) Engineering

strategies for periferal nerve repair orthop. *Clin. North Am.*, **31**, 485–497.

126. Yang, Y., De Laporte, L., Rives, C.B., Jang, J.H., Lin, W.C., Shull, K.R., and Shea, L.D. (2005) Neurotrophin releasing single and multiple lumen nerve conduits. *J. Control. Release*, **104**, 433–446.

127. Sterne, G.D., Brown, R.A., Green, C.J., and Terenghi, G. (1997) Neurotrophin-3 delivered locally via fibronectin mats enhances peripheral nerve regeneration. *Eur. J. Neurosci.*, **9**, 1388–1396.

128. Hontanilla, B., Auba, C., and Gorria, O. (2007) Nerve regeneration through nerve autografts after local administration of brain-derived neurotrophic factor with osmotic pumps. *Neurosurgery*, **61**, 1268–1275.

129. Thompson, B.C., Richardson, R.T., Moulton, S.E., Evans, A.J., O'Leary, S., Clark, G.M., and Wallace, G.G. (2010) Conducting polymers, dual neurotrophins and pulsed electrical stimulation – dramatic effects on neurite outgrowth. *J. Control. Release*, **141** (2), 161–167.

130. Richardson, R.T., Thompson, B., Moulton, S., Newbold, C., Lum, M.G., Cameron, A., Wallace, G.G., Kapsa, R., Clark, G., and O'Leary, S. (2007) The effect of polypyrrole with incorporated neurotrophin-3 on the promotion of neurite outgrowth from auditory neurons. *Biomaterials*, **28**, 513–523.

131. Thompson, B.C., Moulton, S.E., Ding, J., Richardson, R., Cameron, A., O'Leary, S., Wallace, G.G., and Clark, G.M. (2006) Optimising the incorporation and release of a neurotrophic factor using conducting polypyrrole. *J. Control. Release*, **116**, 285–294.

132. Bryan, D.J., Holway, A.H., Wang, K.K., Silva, A.E., Trantolo, D.J., Wise, D., and Summerhayes, I.C. (2000) Influence of glial growth factor and Schwann cells in a bioresorbable guidance channel

on peripheral nerve regeneration. *Tissue Eng.*, **6**, 129–138.

133. Brown, D., Meagher, P., Knight, K., Keramidaris, E., Romeo-Meeuw, R., Penington, A.J., and Morrison, W. (2006) Survival and function of transplanted islet cells on an in vivo, vascularized tissue engineering platform in the rat: a pilot study. *Cell Transplant.*, **15** (4), 319–324.

134. Penington, A.J., Craft, R.O., and Morrison, W.A. (2006) A defined period of sensitivity of an experimental burn wound to a second injury. *J. Burn Care Rehabil.*, **27** (6), 882–888.

135. Penington, A.J., Haloi, A.K., Ditchfield, M., and Phillips, R. (2006) Facial infiltrative lipomatosis. *Pediatr. Radiol.*, **36** (11), 1159–1162.

136. Penington, A.J. and Morrison, W.A. (2007) Skin graft failure is predicted by waist-hip ratio: a marker for metabolic syndrome. *ANZ J. Surg.*, **77**, 118–120.

137. Penington, A.J. (2008) Negative results and the limitations of power. *ANZ J. Surg.*, **78**, 99–102.

138. Luthert, P., Vrbova, G., and Ward, K.M. (2008) Effects of slow frequency electrical stimulation on muscles of dystrophic mice. *J. Neurol. Neurosurg. Psychiatr.*, **43**, 803–809.

139. Bouchentouf, M., Benabdallah, B.F., Mills, P., and Tremblay, J.P. (2006) Exercise improves the success of myoblast transplantation in mdx mice. *Neuromuscul. Disord.*, **16**, 518–529.

140. Stanish, W.D., Valiant, G.A., Bonen, A., and Belcastro, A.N. (1982) The effects of immobilization and of electrical stimulation on muscle glycogen and myofibrillar ATPase. *Can. J. Appl. Sport Sci.*, **7**, 267–271.

141. Goldspink, G. and Yang, S.Y. (2001) Effects of activity on growth factor expression. *Int. J. Sport Nutr. Exerc. Metab.*, **11** (Suppl. 201), S21–S27.

2
Carbon

Carbon provides the framework for all living tissue and plants; without this element, life as we know it would not exist. Many biomedical materials (polymers, polysaccharides, etc.) have carbon as their core component and, as a result, the breadth of literature covering research into these carbon-based materials is immense. Therefore, this chapter focuses only on the allotropic forms of carbon, namely, graphene and carbon nanotubes, and their applications in bionics research.

2.1
Introduction to Carbon

Carbon is one of the building blocks of life. It is present in all known life-forms and is the second most abundant element by mass (about 18.5%), after oxygen, in the human body. Carbon exists in a number of allotropes (different solid forms) commonly known as *diamond, graphite, carbon nanotubes (CNTs)*, or *fullerenes* (Figure 2.1).

All of these carbon allotropes, or composites containing them, have been investigated for use as biomaterials. Targeted areas of applications, including cardiac pacemaker electrodes [1], recording and stimulating electrodes for the central nervous system [2], and drug delivery vehicles [3], have utilized the electrical conductivity and biocompatibility of the carbon allotrope. The mechanical and biocompatible properties have resulted in the use of carbon material in areas such as coatings for implantable orthopedic [4] and cardiovascular prosthetics [5, 6]. In this chapter, we focus on those allotropes that have been investigated for and used in "bionic" applications. Fundamental to the application of carbon in bionics is the appropriate understanding and utilization of a carbon material's biocompatibility, coupled to its ability to deliver and receive electrical signals (electrical stimulation and recording).

Organic Bionics, First Edition. Gordon G. Wallace, Simon E. Moulton, Robert M.I. Kapsa, and Michael J. Higgins.
© 2012 Wiley-VCH Verlag GmbH & Co. KGaA. Published 2012 by Wiley-VCH Verlag GmbH & Co. KGaA.

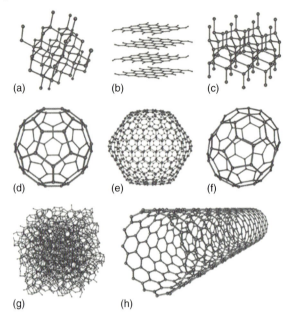

(a)   (b)   (c)

(d)   (e)   (f)

(g)   (h)

Figure 2.1 Various forms of carbon: (a) diamond,
(b) graphite, (c) ionsdaleite, (d–f) fullerenes, (g)
amorphous carbon, and (h) carbon nanotube. (Modified from "Carbon." Chemicool Periodic Table.
http://www.chemicool.com/elements/carbon.html.)

2.2
Graphene

Graphene is the name given to a two-dimensional sheet of sp^2-hybridized
carbon. Its extended honeycomb network is the basic building block of other
important allotropes; for example, it can be stacked to form 3D graphite,
rolled to form 1D nanotubes, and wrapped to form 0D fullerenes. Long-range
π-conjugation in graphene yields extraordinary thermal, mechanical, and
electrical properties, which have long been the interest of many theoretical
studies and, more recently, became an exciting area for experimentalists [7].
In the family of carbon nanostructures, graphene is, experimentally at least,
the youngest member but has attracted enormous recent interest including
the 2010 Nobel Prize in Physics.[1] As with CNTs, graphene is promising
for applications in a wide range of areas such as sensors, high-performance
nanocomposites, and electrodes for optoelectronic and electrochemical devices [8]. Compared to CNTs, the production and processing of graphene
appears to be easier and more economic.

1) *http://nobelprize.org.*

A number of approaches have been proposed to produce graphene sheets, including mechanical cleavage of graphite [8], epitaxial growth on SiC substrates at high temperatures [9], chemical vapor deposition [10, 11], bottom-up organic synthesis [12], solvothermal reaction [13], and chemical conversion of graphite oxide to graphene [14]. Among these methods, the graphite oxide route has received significant attention. This method allows for the large-scale production of graphene at relatively low cost via the reduction of graphene oxide to partially functionalized graphene, which can then be readily solution processed [15]. Graphite is composed of stacks of graphene layers and is abundant in nature and industry. Graphite appears to be an ideal source for large-scale production of graphene. However, high-yield direct exfoliation of graphite has proved to be difficult because of strong intersheet van der Waals interactions. It has recently been demonstrated that sonication of graphite in certain organic solvents or in water in the presence of surfactants can lead to partial exfoliation of graphene sheets [16–18] with limited success and unsatisfactory yields. In contrast, it has long been known that graphite oxide, a layered material that can be produced by controlled oxidation of graphite, can be readily dispersed in water because of the presence of hydrophilic hydroxyl, epoxide, carbonyl, and carboxyl functional groups. Ruoff and coworkers [14] recently confirmed that graphite oxide can be fully exfoliated in water as individual graphene oxide sheets via sonication processing. Although graphite oxide is electrically insulating, it can be conveniently converted back to conducting graphene via chemical reduction or thermal treatment [14, 19]. Rapid thermal expansion can also cause graphite oxide to delaminate into individual graphene sheets [20].

In addition to high yield conversion, solution processability of graphene can be easily achieved through the graphite oxide route. Chemical conversion from graphene oxide leaves a small amount of residual oxygen-containing groups, which make the resulting graphene surfaces negatively charged when dispersed in water. Li *et al.* [19] found that graphene sheets can form stable aqueous colloids via electrostatic repulsion without the need for foreign polymeric or surfactant dispersants. Ruoff and coworkers [21] have observed that stable dispersions can be obtained if the reduction of graphene oxide is conducted in certain organic solvents, such as N,N'-dimethylformamide. Kaner and coworkers [22] have found that graphene oxide can be dispersed in pure liquid hydrazine while being deoxygenated. The resulting graphene can remain dispersed in hydrazine through the formation of a hydrazinium graphene complex in the absence of any surfactants.

The synthesis of CCG (chemically converted graphene) has enabled further processing of graphene into useful structures such as graphene paper (Figure 2.2) that exhibits remarkable mechanical and electrical properties.

Graphene can be rendered soluble in a broad range of solvents through chemical modification. The oxygen-containing groups that exist in graphite oxide (i.e., carboxyl and hydroxyl) can be used to anchor a variety of functional

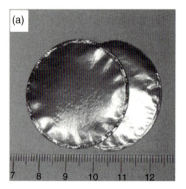 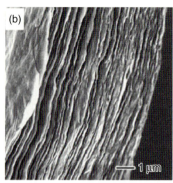

Figure 2.2 (a) Photograph of two pieces of freestanding graphene paper fabricated by vacuum filtration of chemically prepared graphene dispersions, followed by air drying and peeling off the membrane. (b) Side- view scanning electron microscopic (SEM) images of an about 6 mm thick sample at high magnification.(Adapted with permission from Chen *et al.* [23].)

groups. Alkylamines [24], isocyanates [25], and diazonium salts [26] have been attached using this approach. Polymers have also been grafted onto graphene sheets [27].

The aggregation of graphene can also be prevented by adding other stabilizing agents such as surfactants and polymers into the graphene oxide dispersion before deoxygenation [25, 28, 29].

As stated in Chapter 1, what makes a material attractive for bionic applications are its appropriate electronic, electrochemical, chemical, and mechanical properties, as required for bionic materials. These properties are discussed in detail below for graphene. While most of these properties have been investigated with respect to the use of graphene in nonbionic applications, the possibilities for use in bionics are obvious.

2.2.1
Properties of Graphene

2.2.1.1 Electronic Properties
Graphene is quite different from most conventional three-dimensional materials. Intrinsic graphene is a semimetal or a zero-gap semiconductor. The electronic properties of graphene change with the number of layers, and the relative position of atoms in adjacent layers (stacking order) is also important [30]. For bilayer graphene, the stacking order can be either AA, with each atom on top of another atom, or AB, where a set of atoms in the second layer sits on top of the empty center of a hexagon in the first layer.

As the number of layers increases, the stacking order can become more complicated.

As stated above, a graphene crystal is an infinite two-dimensional layer consisting of sp^2-hybridized carbon atoms, which belongs to one of the five 2D Bravais lattices called the *hexagonal (triangular) lattice*. Graphene was initially considered as a theoretical building block that was used to describe the graphite crystal as well as to describe the formation of CNTs (as rolled graphene sheets) and predict their fascinating electronic properties. A comprehensive discussion of the electronic properties of graphene is contained in the review by Terrones *et al.* [30] and references cited therein.

The extraordinary electronic properties of graphene can be attributed to the high quality of its 2D crystal lattice [31–33]. This high quality implies an unusually low density of defects, which typically serve as the scattering centers that inhibit charge transport. In 2008, Kim *et al.* [34] measured carrier mobility in excess of 200 000 cm^2 V^{-1} s^{-1} for a single layer of mechanically exfoliated graphene (Figure 2.3). In their insightful experiments, substrate-induced scattering was minimized by etching under the channel to produce graphene completely suspended between gold contacts. At such high carrier mobility, charge transport is essentially ballistic on the micrometer-scale at room temperature. This has major implications for the semiconductor industry because it enables, in principle, fabrication of all-ballistic devices even at today's integrated circuit (IC) channel lengths (currently down to 45 nm).

The second important point about charge transport in graphene is am-bipolarity. In the field-effect configuration, this implies that carriers can be tuned continuously between holes and electrons by supplying the requisite gate bias. This can be easily visualized, given the unique band structure of graphene [8]. Under negative gate bias, the Fermi level drops below the Dirac point, introducing a significant population of holes into the valence band. Under positive gate bias, the Fermi level rises above the Dirac point, promoting a significant population of electrons into the conduction band. Significantly, the ambipolar charge transport properties of graphene, as well as the ability to tune them, give rise to unique electrochemical properties that will be important to electronic biological interfaces.

2.2.1.2 Electrochemical Properties

Graphene exhibits a wide working electrochemical potential window of up to 2.5 V in 0.1 M phosphate buffer solution (pH 7.0) [35] comparable to graphite, glassy carbon, and boron-doped diamond, which have been commonly used as electrodes in electrochemical studies [35–37]. The charge-transfer resistance on graphene, as determined from AC impedance spectroscopy, is much lower than that observed with graphite or glassy carbon electrodes [35]. The reduced impedance is an attractive property as this suggests that

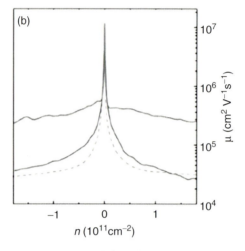

Figure 2.3 Suspended graphene shows extremely high mobility because of the minimization of substrate-induced scattering. (a) SEM image of a suspended sheet after etching. (b) Field-effect measurements indicate mobility $>200\,000\,\text{cm}^2\,\text{V}^{-1}\,\text{s}^{-1}$. (Adapted with permission from Bolotin *et al.* [34].)

graphene electrodes may be more sensitive to recording neural activity as well as having the ability to deliver electrical stimulation in a more efficient manner (i.e., consuming less power).

Cyclic voltammetry (CV) has been used to study the electron-transfer characteristics of graphene. Redox couples, such as $[Fe(CN)_6]^{3-/4-}$ (Figure 2.4) and $[Ru(NH_3)_6]^{3+/2+}$, are reported to exhibit well-defined redox peaks [38–40]. Both anodic and cathodic peak currents in the CVs are linear with the square root of the scan rate, which suggests that the redox processes on graphene-based electrodes are predominantly diffusion controlled [39]. The peak-to-peak potential separations (ΔE_p) have been found to be close to the ideal value of $59\,\text{mV}$, for example, 61.5–$73\,\text{mV}$ ($10\,\text{mV s}^{-1}$) for $[Fe(CN)_6]^{3-/4-}$ redox couple [38, 40–42] and 60–$65\,\text{mV}$ ($100\,\text{mV s}^{-1}$) for $[Ru(NH_3)_6]^{3+/2+}$ system [40]. These ΔE_p values are indicative of a more ideally reversible electron-transfer behavior than are typically observed on a glassy carbon [36]. ΔE_p relates to the electron-transfer rate

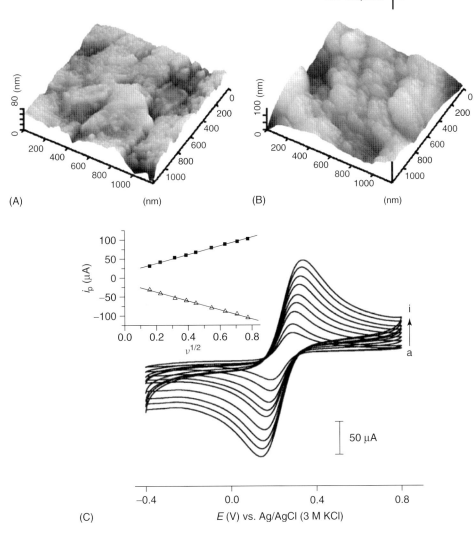

Figure 2.4 AFM images of the graphite substrate (A) before and (B) after modification with graphene. (C) Cyclic voltammograms at the graphene-modified EPPGE in 0.1 M KCl containing 5 mM $Fe(CN)_6^{-4}$ at scan rates of (a) 25, (b) 50, (c) 100, (d) 150, (e) 200, (f) 300, (g) 400, (h) 500, and (i) 600 mV s^{-1}. The inset shows the anodic and cathodic peak currents as a function of the square root of scan rates. (Adapted with permission from Lin et al. [39].)

coefficient [43] and the ΔE_p values obtained, therefore, indicate fast electron transfer for single-electron electrochemical reaction processes [41] on graphene.

Tang et al. [40] systematically studied the behavior of three redox couples: $[Ru(NH_3)_6]^{3+/2+}$, $[Fe(CN)_6]^{3-/4-}$, and $Fe^{3+/2+}$. $[Ru(NH_3)_6]^{3+/2+}$ is a nearly ideal outer-sphere redox system that is insensitive to most surface defects or impurities on electrodes and can serve as a useful benchmark in comparing electron transfers at various carbon electrodes. Other redox couples such as $[Fe(CN)_6]^{3-/4-}$ are "surface sensitive" but not "oxide sensitive," and $Fe^{3+/2+}$ is both "surface sensitive" and "oxide sensitive" [36]. The apparent electron-transfer rate constants (k_0) calculated from cyclic voltammograms for $[Ru(NH_3)_6]^{3+/2+}$ obtained at graphene and glassy carbon electrodes are 0.18 and 0.055 cm s^{-1}, respectively [40]. This indicates that the unique electronic structure of graphene, especially the high density of the electronic states over a wide energy range, endows graphene with fast electron-transfer characteristics [36]. The k_0s for $[Fe(CN)_6]^{3-/4-}$ at graphene and glassy carbon electrodes were calculated to be 0.49 and 0.029 cm s^{-1}, respectively, and the electron-transfer rates for $Fe^{3+/2+}$ at the graphene electrode were several orders of magnitude higher than that at glassy carbon electrodes [40]. These results indicated that the electronic structure and surface physicochemistry of graphene were beneficial for enhancing electron-transfer processes [36, 40, 44].

The superior electron-transfer property of graphene suggests that electrodes based on graphene would be advantageous for detecting biological and biochemical processes and/or for delivery of charge into biological systems as they produced more defined redox responses with respect to conventional glassy carbon electrodes [45]. The edges of multilayer graphene nanoflakes [41], or the edges and functional groups of reduced graphene oxide [46], have been utilized to detect important neurotransmitters such as dopamine and serotonin. More traditional electrodes have been shown to be less sensitive toward these systems with broad, poorly defined redox responses. Shang *et al.* [41] used microwave-plasma-enhanced chemical vapor deposition (CVD) to grow multilayer graphene nanoflake films on Si substrates without the use of any catalysts. The resulting films comprised multilayer nanoflakes with exposed sharp edges that can undergo redox reactions in solution and, therefore, detect electrochemically active biomolecules. This behavior is consistent with that observed on active edge-plane sites on highly oriented pyrolytic graphite (HOPG) [47] and the tips of CNTs [48]. CV measurements with a graphene nanoflake electrode showed clear voltammetric peaks for the direct oxidation of dopamine in solution. Clear peaks were also observed for ascorbic acid (vitamin C) and uric acid, which are potentially interfering molecules, when analyzed in separate solutions. Crucially, these peaks remained well resolved in a mixed solution of the three compounds, thus showing that the nanoflake electrodes could be used to clearly identify dopamine in mixtures. These results contrast with the response of a traditional glassy carbon electrode, in which the biomolecules showed similar broad peaks at higher potentials, whether analyzed individually or in mixtures. Similar effects have been observed previously in electrochemical studies with electrodes that incorporate CNTs [49, 50]. The stark contrast in performance of the two electrodes demonstrated the faster electron-transfer kinetics and favorable electrocatalysis provided by the defects at the edges of the graphene nanoflakes.

Direct electrochemical interrogation of enzymes, that is, the direct electron communication between the electrode and the active center of the enzyme without the participation of mediators or other reagents [51–53], is significant in the development of biosensors, biofuel cells, and biomedical devices [52, 54–56]. Owing to its extraordinary electron transport property and high specific surface area [8], functionalized graphene is expected to promote electron transfer between electrode substrates and enzymes [57]. Shan *et al.* [57] and Kang *et al.* [58] reported the direct electrochemistry of glucose oxidase (GOX) on graphene, where Shan *et al.* [57] employed chemically reduced graphene oxide and Kang *et al.* [58] employed thermally split graphene oxide [59], both exhibiting similar excellent direct electrochemical communication with GOX.

2.2.1.3 Chemical Properties

Highly crystalline graphene surfaces appear to be chemically inert, and these surfaces usually interact with other molecules via physical adsorption ($\pi - \pi$ interactions). However, several chemical groups such as carboxyl (COOH), carbonyl (CO), hydrogenated (CH), and amines (NH_2) found at the edges of graphene sheets are more chemically reactive. The chemical reactivity can change drastically at these edges depending on their carbon termination (i.e., armchair or zigzag). In order to make the graphene surface more chemically reactive, either surface defects or high degrees of curvature need to be introduced.

Oxidation of graphene edges followed by chemical functionalization of the induced defects assists in the formation of stable dispersions in aqueous solvents by preventing destabilizing agglomeration in solution [60]. The functional groups attached to graphene can be small molecules, as described below [59, 61, 62], or polymer chains [63–65]. The chemical functionalization of graphene is a particularly attractive aim because it can improve the solubility and processability, as well as enhance the interactions with organic polymers [24, 66]. Considerable work has been carried out on amination, esterification [20], isocyanate modification [64], and polymer wrapping [63, 64] as routes for the functionalization of graphene. The electrochemical modification of graphene using ionic liquids has also been reported [67].

Fluorinated graphite has been synthesized for decades. However, graphene has been recently fluorinated using different techniques, including plasma treatments and high temperature treatments [68, 69]. Fluorinated graphene can be dispersed homogeneously in solvents and could therefore be used to produce polymer composites [70]. A fully hydrogenated graphene sheet, termed *graphane*, was predicted by Sofo *et al.* [71] and reported experimentally by Elias and co-workers [72]. It has been shown that the electronic properties of graphene can be changed by passivating its surface followed by desorption of hydrogen in specific patterns [73].

2.2.1.4 Mechanical Properties

The mechanical strength of individual graphene sheets has been studied by Hone and coworkers [74], who demonstrated that graphene has a breaking strength that is 200 times larger than that of steel, with a Young's modulus of about 1 TPa. These measurements appear to depend on the number and types of defects present within the sheet. Nanoindentation techniques are well suited for measuring the macroscopic mechanical properties of graphene, including Young's modulus and bending stiffness [75]. For example, by using nanoindentation methods on suspended multilayer graphene flakes, the bending stiffness has been measured and found to be in the range from 2×10^{-14} to 2×10^{-11} N m^{-1} for 8–100 layers, respectively. Static nanoindentation experiments based on the deflection of atomic force microscope (AFM) cantilevers pressed within 100 nm of the center of ~1 mm long double-clamped graphene films provided a measurement of the effective spring constant of multilayer graphene (1–5 N m^{-1}). The spring constant was found to scale with the dimensions of the suspended region and layer thickness (from 5 to 30 layers), as also the extracted Young's modulus of 0.5 TPa although independent of thickness [76]. Other studies have reported a strength of ~40 N m^{-1} and a Young's modulus of ~1.0 TPa, reaching the theoretical limit [74]. Graphene can also be stretched elastically by as much as 20%, more than any other crystal [74].

A significant limitation of the use of nanoindentation techniques is the requirement of a graphene layer to be suspended. The presence of a substrate over which graphene may be either deposited (SiO$_2$[77], glass, sapphire [78], or polymers [79, 80]) or directly grown epitaxially (SiC [81, 82] and metals [83, 84]) makes it hard to separate the intrinsic mechanical properties of graphene from that of the substrate. In contrast to nanoindentation, Raman spectroscopy provides access to information relating to the underlying chemical bonds. Besides complementing the coarse grained approach of macroscopic elasticity, the interrogation of bond vibrations by optical spectroscopy enables the retrieval of information about mechanical and structural properties of films that can have a monolayer thickness and be strongly interacting with a substrate. Raman spectroscopy has thus been used to measure the mechanical properties of graphene films, both freestanding and on a substrate [80, 85] at room and at elevated temperatures [86, 87].

2.3
Carbon Nanotubes

CNTs with unprecedented mechanical [88] and transport properties [89] have generated global scientific and industrial activity. Their wide electrochemical potential window, chemical stability, and large surface area give them very attractive properties as electrode materials in energy storage [90, 91](capacitors and batteries) and energy conversion (fuel cells [92] and electromechanical actuators [93]). Recent advances in combining soft biological materials with CNTs have provided an extra dimension in enabling CNT-based electrodes to interface with biological systems.

CNTs can be considered to be rolled-up sheets of graphene (Figure 2.5) with diameters in the nanodomain and lengths in the micrometer to millimeter domain; extraordinary aspect ratios are attainable. The structure of a CNT can be conceptualized by wrapping a one-atom-thick layer of graphite called *graphene* into a seamless cylinder. The way the graphene sheet is wrapped is represented by a pair of indices (n,m) called the *chiral vector*. The integers n and m denote the number of unit vectors along two directions in the honeycomb crystal lattice of graphene. If $m = 0$, the nanotubes are called *zigzag*. If $n = m$, the nanotubes are called *armchair* (Figure 2.5); otherwise, they are called *chiral*. Because of the symmetry and unique electronic structure of graphene, the structure of a CNT strongly affects its electrical properties. For a given (n,m) nanotube, if $n = m$, the nanotube is metallic; if $n - m$ is a multiple of 3, then the nanotube is semiconducting with a very small band gap; otherwise, the nanotube

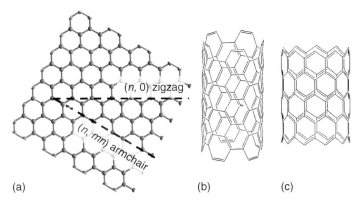

(a) (b) (c)

Figure 2.5 (a) The (n,m) nanotube naming scheme can be thought of as a vector in an infinite graphene sheet that describes how to "roll up" the graphene sheet to make the single-walled carbon nanotubes in (b) armchair and (c) zigzag.

is a moderate semiconductor. Thus, all armchair ($n = m$) nanotubes are metallic, and nanotubes (6,4), (9,1), and so on, are semiconducting.

In the simplest form they form single sheet cylinders, referred to as *single-walled carbon nanotubes (SWNTs)*. SWNTs comprise a seamless cylinder of one atomic layer of carbon consisting of a mixture of sp^3 and sp^2 bonds with the sp^3/sp^2 ratio varying from 15 to 40%. SWNTs are an important variety of CNTs because they exhibit electric properties that are not shared by the multiwalled carbon nanotube (MWNT) variants. SWNTs are the most likely candidates for miniaturizing electronics beyond the microelectromechanical scale currently used in electronics.

The other major form of the CNT is the MWNT which consists of multiple concentric cylinders of increasing diameter about a common axis separated by 0.34 nm, which is the same interplane distance observed in graphite. MWNTs were first produced by Iijima [94] using the arc evaporation process. SWNTs were initially produced using the arc discharge process, wherein the graphite electrodes were modified by transition metal catalysts [95]. SWNTs and MWNTs have also been produced using laser ablation [96] and CVD [97]. Double-walled carbon nanotubes (DWNTs) are a unique form of MWNT that have morphology and physical properties similar to SWNTs but their resistance to chemicals is significantly improved. A DWNT consists of exactly two concentric CNTs; compared to SWNTs, DWNTs have higher mechanical strength and thermal stability and also possess interesting electronic and optical properties [98].

2.3.1
Synthesis

The three main synthesis techniques for CNT production are arc discharge [99], laser ablation [100], and CVD [101]. They differ in the ability to control the quantity and quality of the CNTs, and also in the substrates on which they can be collected. For example, the former two methods use a solid-state carbon precursor for nanotube growth and produce only entangled nanotubes. The use of a hydrocarbon gas as carbon source and metal particles as catalysts in the CVD method enables the production of regularly patterned and highly oriented CNT arrays. All of these methods, however, produce a heterogeneous mixture of metallic and semiconducting CNTs with varying degrees of amorphous carbon impurities and catalyst residues. The removal of these impurities and separation based on their metallic characteristics by techniques such as electrophoretic and chromatographic methods have proved to be quite challenging and produce only limited quantities of purified material [102, 103]. Contrasting these approaches, density-gradient ultracentrifugation is proving to be a scalable process to separate CNTs based on their electronic structure [104].

In terms of CNT processability, most of the available nanotube raw materials are in the form of intractable carbon soot that spontaneously aggregates in both aqueous and organic environments. Many recent studies are devoted to improving their solvent compatibility. Covalent or noncovalent attachment of functional molecules (e.g., polymers, surfactants, biopolymers, zwitterions, etc.) has been found to be essential to produce stable (agglomerate-free) CNT dispersions [105–108]. These dispersions can then be used to fabricate useful electrode structures.

A variety of surfactants (including Triton X-100, sodium dodecylsulfate, and sodium dodecylbenzenesulfonate) have been used to assist CNT dispersion in water. This widely used process involves high-power sonication to break the CNT bundles. The surfactant is believed to be adsorbed onto the CNT surface (randomly or through micelle formation), which facilitates CNT stabilization in aqueous environment. Owing to the batch-to-batch variation of CNT production, many parameters in this procedure (including sonication duration and power, and nanotube-to-surfactant concentration ratios) require continued optimization. Inefficient dispersion results in the formation of aggregates, which compromises the quality of the material (e.g., mechanical properties of a composite) and reproducibility of the results.

2.3.2
Electronic Properties of Carbon Nanotubes

The electronic band structures of graphene and SWNTs are very similar. For graphene and metallic SWNTs, the valence band and the conduction band touch at specific points in the reciprocal space. For semiconducting SWNTs, the conduction band and the valence band do not touch. Semiconducting SWNTs have been extensively studied as channels in transistor devices [109, 110], while metallic SWNTs have been considered for applications such as IC interconnects and field emission [111]. As a one-dimensional object, the conductivity of an SWNT contacted at both ends can be calculated using the Landauer equation (Eq. (2.1)) [109, 112, 113]:

$$G = 2\frac{e^2}{h} \sum_i T_i \tag{2.1}$$

where $2e^2/h$ is the quantum conductance and $\sum_i T_i$ is the sum of the transmission probabilities (the sum of the contributing conduction channels) whose energy is between the electrochemical potential of the two contacts. For a metallic SWNT electrically contacted at both ends, in the absence of scattering, all T_i terms are equal to 1 and the resistance of an SWNT is then $R = 2R_Q (\sim 13 \text{ k}\Omega)$ as an SWNT has two conduction channels. The quantum resistance $R_Q = 6.5 \text{ k}\Omega$ is due to the mismatch between the number of conduction channels in the nanotube and the macroscopic metallic contacts. Theoretical studies [114, 115] have indicated that ideal nanotubes are ballistic conductors for distances in the order of a micrometer. The one-dimensional confinement of electrons combined with the requirements for energy and momentum conservation leads to ballistic conduction. To this ideal resistance, we need to add the scattering resistance due to impurities or nanotube defects, which reduces the mean free path compared to an ideal SWNT (typically $l_0 = 1\,\mu m$).

If the nanotube length (l) is less than $1\,\mu m$, we can usually neglect the scattering resistance. Once we contact the nanotube to a substrate or to a metallic contact, we need to consider an additional contact resistance, which strongly depends on the material in contact with the nanotube and the difference between the work functions of the SWNT and of the electrode material. Shiraishi and Ata [116] found the work functions of MWNTs and SWNTs to be 4.95 and 5.10 eV, respectively. Mann *et al.* [117] demonstrated that highly reproducible ohmic contacts were obtained by contacting Pd–Pbs to metallic SWNTs. They also indicated that Pd contacts were more reliable than Ti contacts and that nonohmic behavior was observed when CNTs were connected to Pt contacts.

Assuming that we can neglect coupling between adjacent CNTs, which is usually a valid assumption as the electrons would rather travel through the ballistic path than through a large tunneling resistance ($2-140$ MΩ) [118], the resistance of a bundle of SWNTs (Eq. (2.2)) can be viewed as a parallel circuit of the resistances of the single nanotubes. If we have nSWNTs, the resistance of the bundle will be

$$R_{\text{SWNT bundle}} = \frac{R_{\text{SWNT}}}{n} \tag{2.2}$$

The electrical properties of SWNTs have been extensively studied [119–122], often in the context of developing devices [93] such as interconnects [123–125] or CNT-based transistors [109, 126, 127]. In contrast, the electrical properties of MWNTs [128, 129] have not been investigated at the same level of detail because of the additional complexities arising from their structure, as every shell has different electronic characteristics and chirality in addition to interactions between the shells [130, 131].

However, for MWNTs connected at both ends by metallic contacts, the electronic transport is dominated by outer-shell conduction at low bias and temperature [89, 132, 133]. Theoretical models [124] and experimental results [134] point to the critical role of shell-to-shell interactions in significantly lowering the resistance of MWNTs with a large number of walls.

2.3.3
Electrochemistry of Carbon Nanotubes

Compton and coworkers [135] compared the redox properties of NADH, epinephrine, and norepinephrine at MWNTs and graphite microparticle-modified basal-plane pyrolytic graphite (BPPG) electrodes.

They found that the response at MWNTs and graphite-modified BPPG is essentially the same since the edge-plane sites and the tube ends provide the same electrochemically active sites [47, 136]. The pristine walls of CNTs behave as a basal plane of graphite [137], while the ends of the CNTs and the defects on their walls resemble the behavior and fast electron transfer of edge-plane pyrolytic graphite (EPPG) [47, 138]. It is possible to tune the electrochemical activity of MWNTs [139, 140].

Holloway *et al.* [141] compared the response of pristine close-ended SWNTs (i.e., containing no "edge-like" sites) annealed in vacuum with SWNTs that were consequently electrochemically activated in a manner similar to that reported by Pumera *et al.* [139] and Musameh *et al.* [142]. These studies indicated that the walls of SWNTs are not electrochemically active.

An important experiment to demonstrate the role of the ends and walls of nanotubes (Figure 2.6) was carried out by Dai and coworkers [143]. The electrochemical properties of the ends (CNT-T) and walls (CNT-S) of CNTs were compared in a way that the walls or ends were selectively masked with a nonconductive polymer coating (Figure 2.6a). They found that the electrochemistry of several analytes was sensitive to the exposed CNT region (thus sensitive to the ends/walls) and to the oxidation state (Figure 2.6b) of such a region. This is an important addition to previous knowledge, but it does not contradict it, because the walls of the CNTs used in this study were not free of defects and thus they contained "edge-like" defects, which were previously shown to be electrochemically active.

2.3.4
Chemical and Biological Properties of Carbon Nanotubes

The chemical modification of CNTs is an emerging area in materials science with many bionic applications of CNTs relying on successful surface functionalization.

Arising from their demonstrated ability to cross membrane barriers, CNTs [144] are of great potential value to biomedical applications. The translocation and biodistribution of pristine CNTs in mice [145–147] and plants [148, 149] have been studied by several groups. However, the great potential value of CNTs in biomedical applications is counterbalanced by contradictory data demonstrating toxic effects of SWNTs [150–154]. This strongly suggests the need for a standardized framework to understand the bioactivities (as a function of surface chemistry) of CNTs and assess their toxicities *in vivo* [155]. A more detailed discussion of the toxicity of CNTs is presented in Chapter 4.

Recent interest in introducing CNTs into biological systems has led to the use of biopolymers (peptides, polysaccharides, DNA, etc.) to facilitate CNT stabilization in aqueous media [156–158]. Some amphiphilic α-helical peptides have been specifically designed to maximize their affinity for CNT

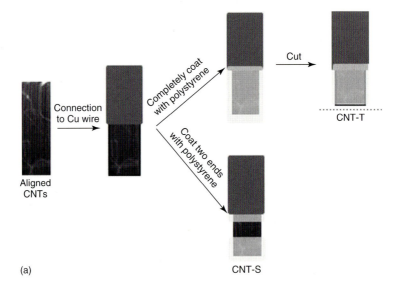

(a)

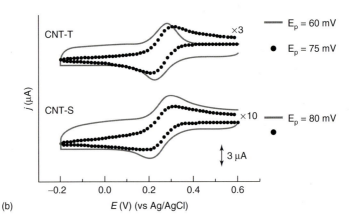

(b)

Figure 2.6 (a) Schematic representation of the procedure for preparing the CNT electrodes with only the nanotube tip (CNT-T) or sidewall (CNT-S) accessible to electrolyte. (b) Cyclic voltammograms of 5 mM $K_3[Fe(CN)_6]$ recorded in 0.1 M PBS (pH 6.5) at a scan rate of 100 mV s^{-1} for the CNT-T (upper dotted curve), oxidized-CNT-T (upper solid curve), CNT-S (lower dotted curve), and the oxidized-CNT-S electrodes (lower solid curve). For clarification, the current responses at the CNT-T and CNT-S electrodes were amplified and the amplification coefficients are also shown. (Adapted with permission from Gong *et al.* [143].)

Table 2.1   Examples of covalent attachment of functional groups to CNTs.

SWNT	Functional group	References
	Carboxylic acid	[161–163]
	Ester	[164]
	Amines	[163, 165–167]
	Thiols	[168]
	Proteins	[169, 170]
	DNA	[171–173]
	Enzymes	[174]
	Glucose	[175]
	Alkylation	[176]
	Fluorination	[177]
	Nucleophilic carbenes	[178]
	Phenyl groups	[179]
MWNT	Carboxylic acid	[180, 181]
	Polyethylenimine	[182]

surfaces through $\pi-\pi$ stacking interactions. The same principle allows SWNT bundles to be solubilized by certain conjugated polymers including poly(*m*-phenylenevinylene) and its derivatives [159, 160]. Both covalent (chemical bonds, Table 2.1) and noncovalent (electrostatic, van der Waals, and/or hydrophobic interactions, Table 2.2) attachment of CNTs have been explored.

Owing to our increased understanding of the surface chemistry of CNTs and their fundamental interactions, several general functionalization strategies for the noncovalent and covalent attachment of biomolecules have been envisaged [205]. These include the covalent coupling typically via carboxylic acid groups present at defects sites and open ends, functionalization via surfactants and wetting layers, physical wrapping of proteins, and insertion of molecules into their internal cavities (Figure 2.7). A full complement of biomolecules, including DNA, peptides, enzymes, antibodies, and polysaccharides, have been immobilized and have been well reviewed in the literature [205]. These exciting developments provide a background for unprecedented future functionalities that may be introduced into CNT-based medical bionic structures when biosafety issues have been fully defined and assimilated into therapeutic application requirements [206]. Details of manipulating the CNT–biomolecule interactions via commonly used chemical functionalization routes are also well studied [207]. Here, we narrow the focus to composite biomaterials and composites where the electrical properties play a main role in the biofunctionalized CNT composite such as in biosensors and stimulating/recording electrodes (e.g., neural electrodes).

Table 2.2 Examples of noncovalent attachment of functional groups to CNTs.

SWNT	Functional group	References
	Sodium dodecylsulfonate	[183]
	Triton X-100	[184]
	Sodium dodecylbenzene sulfonate	[106, 185]
	Porphyrin	[186]
	Poly(metaphenylenevinylene)	[187, 188]
	Poly(vinyl alcohol)	[189] (and references therein) [190]
	Nafion	[191]
	DNA	[192, 193]
	Proteins	[169, 193–198]
	Peptides	[199]
	Enzymes	[200, 201]
	Polysaccharide	[202]
MWNT	Polylactic acid	[203]
	DNA	[193, 204]
	Chitosan, hyaluronic acid	[204]
	Proteins	[195]

Control of biocompatibility and nanotoxicity of CNTs has also been achieved by modifying surface chemistry. This rich surface chemistry has resulted in significant recent progress in using CNTs as effective carriers for delivering various peptides, nucleic acids, antigens, and small molecular drugs into living cells [208–211]. Dai and coworkers complexed SWNTs with a large amount of doxorubicin (DOX, a popular anticancer drug), where the complexation was due to noncovalent attachment via $\pi-\pi$ stacking of the DOX aromatic hydroxyl-anthraquinonic rings to the nanotube surface [212]. According to the *in vitro* toxicity experiments, the DOX-loaded phospholipid (PL)-SWNT induced significant U87 cancer cell death and cell apoptosis compared to PL-SWNT.

The surface chemistry can also be tailored to deliver large functional biomolecules into cells. Pantarotto *et al.* [213] covalently linked a neutralizing B cell epitope from the foot-and-mouth disease virus (FMDV) to mono- and bis-derivatized CNTs. Immunological characterization of these conjugates revealed that the epitope was appropriately presented after conjugation to CNTs for recognition by antibodies as measured by BIAcore technology. Moreover, peptide-CNTs elicited strong antipeptide antibody responses in mice with no detectable cross-reactivity to the CNTs. However, only the mono-derivitized CNT conjugate induced high levels of virus-neutralizing antibodies. These findings demonstrated for the first time the potential of

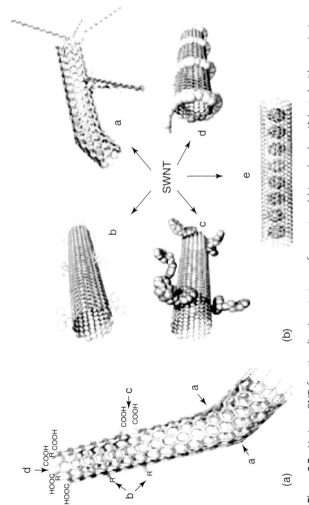

Figure 2.7 Various CNT functionalization strategies for guest and biomolecules. (Adapted with permission from Katz et al. [205].)

CNTs to present biologically important epitopes in an appropriate conformation both *in vitro* and *in vivo* [214]. Zhang *et al.* [215] conjugated siRNA with SWNTs and specifically targeted murine telomerase reverse transcriptase (mTERT) expression to form the mTERT-siRNA:SWNT + complex. These functionalized SWNTs successfully entered three cultured murine tumor cell lines, silenced the expression of the targeted gene, inhibited cell proliferation, promoted cell senescence *in vitro*, and suppressed the tumor *in vivo*. Similarly, Dai and coworkers [216] employed SWNTs as molecular transporters to deliver siRNA into human T cells and primary cells. It was found that SWNT conjugates have superior silencing effects over conventional liposome-based nonviral agents.

Functionalization of CNTs also renders them more amenable to processing into useful structures. Flavin adenine dinucleotide (FAD), the redox-active prosthetic group of flavoenzymes that catalyze important biological redox reactions, and the flavoenzyme GOX were both found to spontaneously adsorb onto CNT bundles [217]. For both constituents (GOX and FAD), quasi-reversible one-electron transfer to the CNT was observed, highlighting the potential for reagentless biosensors. In contrast to mesoscale CNT structures (e.g., bundles), individual SWNTs have been GOX functionalized by use of a linking molecule, which on one side binds to the SWNT through noncovalent, van der Waals coupling with a pyrene group and on the other side covalently binds the enzyme through an amide bond. For these single SWNT/GOX devices, measurements of conductance versus pH showed that on changing the pH from 4 to 5.5 the conductance decreased by 0.3 µS, indicating their excellent ability to measure pH changes down to 0.1 [218]. Lin *et al.* [219] have demonstrated the operation of biosensors in real-tissue environments by integrating CNT/glucose dehydrogenase (GDH) and CNT/l-lactate dehydrogenase (LDH) with microelectrode sensors implanted into rat brain tissue to detect glucose and lactate following cerebral ischemia and reperfusion. The methods they used involved incorporating the enzymes into a CNT/methylene green (MG) composite by dip coating the CNT/MG into a mixture of 1% BSA, 1% glutaraldehyde, and enzyme constituent, presumably to noncovalently adsorb onto BSA followed by covalent attachment of the enzyme via glutaraldehyde as a linker. On-line continuous and simultaneous monitoring using the implanted microsensor was successful in recording the basal levels of glucose and lactate (0.39 ± 0.03 and 0.93 ± 0.05 mM, respectively) and their subsequent changes associated with an inadequate supply of blood/oxygen to the brain due to ischemic stroke [219].

One of the major considerations in this area is increasing the transfer speed of electrons between the enzyme and electrochemical transducer, which has been greatly improved through site-selective covalent immobilization. Here, the difficulty is not in the attachment of the enzyme to the CNT but rather the configuration of the CNT bulk assembly in the biosensor [220]. To address this, CNT/GOX electrodes have been implemented by firstly modifying a gold substrate with cysteamine and then by aligning the SWNT perpendicular to the electrode using self-assembly [221]. The CNT/GOX was prepared by covalent attachment of GOX either to the ends of the CNT or noncovalently reconstituted around the CNT. Importantly, both approaches provided advantages, with the latter shown to provide more efficient electron transfer (rate constant 9 s^{-1}) to the redox-active site, FAD (Figure 2.8a). Elaborate sensing devices have been achieved with covalent coupling of primary antibodies to CNT forests to facilitate secondary binding of a complementary antibody–MWNT–horseradish peroxidase (HRP) conjugate that significantly increases the amount of the electrochemical label, HRP, and electrochemical signal generated [222] (Figure 2.8b). Furthermore, the device could be used to directly measure prostate-specific antigen in samples of human serum from patients and was more sensitive to PSA compared to current commercial clinical immunoassays. Streptavidin has been covalently coupled to CNT using poly(ethylene)-vinylacetate as a linker to provide a functionalized platform for further binding of molecules associated with electrochemical-luminescence detection [223] (Figure 2.8c). An attractive feature of these CNT/streptavidin composites is their applicability for hosting a range of other biotinylated molecules. Some of the various chemical functionalization methods used in the above studies on biosensors are summarized in Figure 2.8.

To enhance the interface between CNT and nerve cells, earlier studies have coated MWNT with bioactive 4-hydroxynonenal (4-HNE), a highly reactive aldehyde formed by lipid peroxidation in cells, via noncovalent adsorption. Even though 4-HNE is generally reported to inhibit neurite outgrowth, rat brain neurons cultured on these composites exhibited increased neurite growth and branching compared to pristine MWNT substrates [224]. Nguyen-Vu et al. [225] have prepared a multicomposite CNT 3D structure by applying a thin layer of type IV collagen adsorbed onto vertically aligned MWNT to promote the adhesion of PC12 nerve cells. Postadsorption of NGF (nerve growth factor) was additionally carried out to induce neurite outgrowth. In addition, electrochemical addition of a polypyrrole (PPy) coating had the advantage of (i) preventing the collapse of the CNT structure, (ii) improving the mechanical contact with cells, and (iii) decreasing the electrode impedance (Figure 2.9).

Carbon-vapor-deposited CNTs containing carboxyl derivatives have been used to provide a covalent anchor for NGF or BDNF (brain-derived neurotrophic factor) via carbodiimide chemistry [226]. In contrast to previous fixed CNT composite film structures, these CNT/NGF composites were endogenously introduced into the culture medium of embryonic DRG (dorsal root ganglion) neurons and shown to induce differentiation. Implementing CNT composites with nerve growth factors in this way is intriguing, particularly if some advantage could be gained from their conductive properties. However, further work is required to clarify the cell

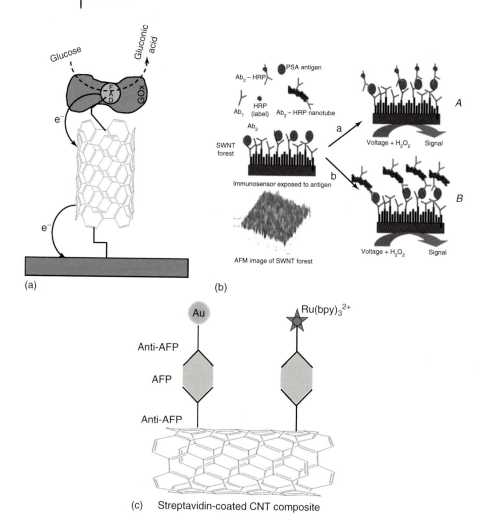

Figure 2.8 (A) Schematic of direct electron transfer through the transducer and redox center (FAD) of GOX via CNT. (B) Construction of sensors made from primary antibodies (Ab$_1$), secondary antibodies (Ab$_2$), horseradish peroxidase (HRP), and a forest of CNT on graphite. Sensor using Ab$_2$–MWNT–HRP bioconjugate. (Reproduced from Yu *et al.* [222].) (C) CNT as an immobilization platform for electrochemical-luminescence sensing.

uptake of these CNT composites and its presently conflicting consequences. Recently, Thompson *et al.* [227] demonstrated the use of an aligned CNT array membrane electrode as a nanostructured supporting platform for PPy films (Figure 2.10), which exhibited significant improvement in controlled release of neurotrophins.

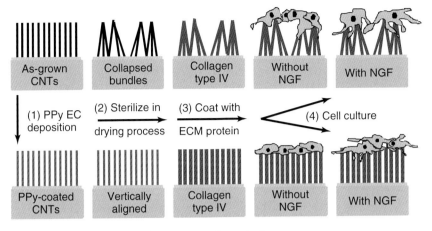

Figure 2.9 Schematic of sample preparation for vertically aligned CNT/PPy/collagen/NGF composites.(Reproduced from Nguyen-Vu et al. [225].)

2.3.5
Mechanical Properties of Carbon Nanotubes

It is well known that CNTs, possessing the cylindrical structure made of a seamless graphene sheet, are expected to have outstanding mechanical properties such as high tensile strength and a high elastic modulus [228]. These properties are the reasons for considering CNTs to be an ideal filler material for composite materials used in aerospace structural elements and, more recently, biomedical applications (i.e., bone regeneration) [229–231].

Theoretical prediction of the mechanical properties of CNTs was first performed by Tomanek et al. in 1993 [232]. By using a Keating potential, the authors have shown that Young's modulus of small SWNTs can be up to 1.5 TPa, which surpasses materials well known for a high tensile strength, such as steel strings and synthetic fibers. Other groups have used different empirical and nonempirical methods, and larger models and multiple layers, to predict various mechanical properties of CNTs [233–236], which also show that CNTs are indeed expected to be resilient materials.

Experimentally, the mechanical properties of CNTs were determined by observing the vibrations of CNTs with a transmission electron microscope (TEM) [237]. By measuring the amplitude of the vibrating MWNTs, Treacy et al. have shown that Young's modulus is about 1.8 TPa. Wong et al. [238] have used an AFM tip with anchored MWNTs to measure the bending force of the CNTs and they obtained an average value of 1.28 TPa. Other experiments have shown similar or worse results depending on the sample

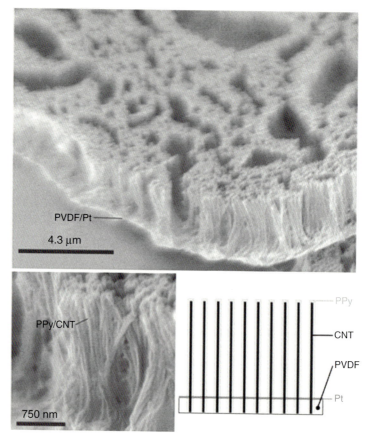

Figure 2.10 SEM images and schematic of composite aligned CNT/PPy/Pt/PVDF material with NT-3 incorporated into the PPy layer.(Reproduced with permission from Thompson *et al.* [227].)

and the experimental conditions [239–241]. The reasons for these deviations in the experimental values are the experimental method itself and the nature of CNTs. Since CNTs are nanosized materials, it is extremely difficult to manipulate and measure their mechanical properties. This makes it hard to reproduce the same experimental conditions for mechanical testing. The microstructure of CNTs differs to a great extent depending on the CNT synthesis method and the synthesis conditions. Even in the same batch of synthesized CNTs, length, layer number, diameter, and crystallinity are often variable. In particular, the crystallinity of the tube wall greatly affects the mechanical properties of CNTs. Therefore, selecting different types of CNTs also affects the experimental values, which results in the large discrepancy among reported values.

From the wide distribution of Young's moduli in various measurements, we can understand the threshold values, which can give an indication of what may be expected from CNTs incorporated into composite materials. Nevertheless, it is well established that incorporation of nanotubes can be used to effectively modulate a material's strength, and the remarkable mechanical and biological properties of CNTs have encouraged widespread investigations into their further applications in biomedical devices. The use of CNTs in composite materials for bionic applications is discussed in more detail in Chapter 4.

2.4
Summary

The inherent biocompatibility of carbon-based materials has led to their use as coatings for a number of biomedical implants. However, the more recent discoveries pertaining to nanostructured carbon materials with extraordinary electronic and mechanical properties and chemistries are yet to be fully exploited. These materials, in combination with other biomaterials and/or other organic conductors such as conducting polymers, provide an excellent platform for developing bionic devices.

References

1. Masini, M., Lazzari, M., Lorenzoni, R., Domicelli, A.M., Micheletti, A., Dianda, R., and Masini, G. (2006) Activated pyrolytic carbon tip pacing leads: an alternative to steroid-eluting pacing leads? *PACE Pacing Clin. Electrophysiol.*, **19** (11), 1832–1835.

2. Inagaki, K., Heiney, S.A., and Blazquez, P.M. (2009) Method for the construction and use of carbon fiber multibarrel electrodes for deep brain recordings in the alert animal. *J. Neurosci. Methods*, **178**, 255–262.

3. Partha, R., Mitchell, L.R., Lyon, J.L., Joshi, P.P., and Conyers, J.L. (2008) Buckysomes: fullerene-based nanocarriers for hydrophobic molecule delivery. *ACS Nano*, **2** (9), 1950–1958.

4. Dowling, D.P., Kola, P.V., Donnelly, K., Kelly, T.C., Brumitt, K., Lloyd, L., Eloy, R., Therin, M., and Weill, N. (1997) Evaluation of diamond-like carbon-coated orthopaedic implants. *Diamond Relat. Mater.*, **6** (2), 390–392.

5. Ripart, A. and Mugica, J. (1983) Electrode-heart interface: definition of the ideal electrode. *PACE Pacing Clin. Electrophysiol.*, **6**, 410–421.

6. Mugica, J., Henry, L., Attuel, P., Lazarus, B., and Duconge, R. (1986) Clinical experience with 910 carbon tip leads: comparison with polished platinum leads. *PACE Pacing Clin. Electrophysiol.*, **9**, 1230–1238.

7. Allen, M.J., Tung, V.C., and Kaner, R.B. (2010) Honeycomb Carbon: A Review of Graphene. *Chem. Rev.*, **110**, 132–145.

8. Geim, A.K. and Novoselov, K.S. (2007) The rise of graphene. *Nat. Mater.*, **6**, 183–191.

9. Berger, C., Song, Z., Li, X., Wu, X., Brown, N., Naud, C., Mayou, D., Li, T., Hass, J., Marchenkov, A.N., Conrad, E.H., First, P.N., and

de Heer, W.A. (2006) Electronic confinement and coherence in patterned epitaxial graphene. *Science*, **3012**, 1191–1192.

10. Kim, K.S., Zhao, Y., Jang, H., Lee, S.Y., Kim, J.M., Kim, K.S., Ahn, J.H., Kim, P., Choi, J.Y., and Hong, B.H. (2009) Large-scale pattern growth of graphene films for stretchable transparent electrodes. *Nature*, **457**, 706–710.

11. Jia, R.X., Ho, J., Nezich, D., Son, H., Bulovic, V., Dresselhaus, M.S., and Kong, J. (2009) Large area, few-layer graphene films on arbitrary substrates by chemical vapor deposition. *Nano Lett.*, **9**, 30–35.

12. Wu, J., Pisula, W., and Muellen, K. (2007) Graphene molecules as potential material for electronics. *Chem. Rev.*, **107**, 718–747.

13. Choucair, M., Thordarson, P., and Stride, J.A. (2009) Gram-scale production of graphene based on solvothermal synthesis and sonication. *Nat. Nanotechnol.*, **4**, 30–33.

14. Stankovich, S., Dikin, D.A., Piner, R.D., Kohlhaas, K.A., Kleinhammes, A., Jia, Y., Wu, Y., Nguyen, S.T., and Ruoff, R.S. (2007) Synthesis of graphene-based nanosheets via chemical reduction of exfoliated graphite oxide. *Carbon*, **45**, 1558–1565.

15. Li, D. and Kaner, R.B. (2008) Materials science. Graphene-based materials. *Science*, **320**, 1170–1171.

16. Blake, P., Brimicombe, P.D., Nair, R.R., Booth, T.J., Jiang, D., Schedin, F., Ponomarenko, L.A., Morozov, S.V., Gleeson, H.F., Hill, E.W., Geim, A.K., and Novoselov, K.S. (2008) Graphene-based liquid crystal device. *Nano Lett.*, **8**, 1704–1704.

17. Hernandez, Y., Nicolosi, V., Lotya, M., Blighe, F.M., Sun, Z., De, S., McGovern, I.T., Holland, B., Byrne, M., Gun'Ko, Y.K., Boland, J.J. *et al.* (2008) High-yield production of graphene by liquid-phase exfoliation of graphite. *Nat. Nanotech.*, **3**, 563–568.

18. Lotya, M., Hernandez, Y., King, P.J., Smith, R.J., Nicolosi, V., Karlsson, L.S., Blighe, F.M., De, S., Wang, Z., McGovern, I.T., Duesberg, G.S., and Coleman, J.N. (2009) Liquid phase production of graphene by exfoliation of graphite in surfactant/water solutions. *J. Am. Chem. Soc.*, **131**, 3611–3620.

19. Li, D., Muller, M.B., Gilje, S., Kaner, R.B., and Wallace, G.G. (2008) Processable aqueous dispersions of graphene nanosheets. *Nat. Nanotech.*, **3** (2), 101–105.

20. McAllister, M.J., Li, J.L., Adamson, D.H., Schniepp, H.C., Abdala, A.A., Liu, J., Herrera-Alonso, M., Milius, D.L., Car, R., Prud'homme, R.K., and Aksay, I.A. (2007) Single sheet functionalized graphene by oxidation and thermal expansion of graphite. *Chem. Mater.*, **19**, 4396–4404.

21. Park, S., An, J., Jung, I., Piner, R.D., An, S.J., Li, X., Velamakanni, A., and Ruoff, R.S. (2009) Colloidal suspensions of highly reduced graphene oxide in a wide variety of organic solvents. *Nano Lett.*, **9**, 1593–1597.

22. Tung, V.C., Allen, M.J., Yang, Y., and Kaner, R.B. (2009) High-throughput solution processing of large-scale graphene. *Nat. Nanotechnol.*, **4**, 25–29.

23. Chen, H., Muller, M.B., Gilmore, K.J., Wallace, G.G., and Li, D. (2008) Mechanically strong, electrically conductive, and biocompatible graphene paper. *Adv. Mater.*, **20** (18), 3557–3561.

24. Niyogi, S., Bekyarova, E., Itkis, M.E., McWilliams, J.L., Hamon, M.A., and Haddon, R.C. (2006) Solution properties of graphite and graphene. *J. Am. Chem. Soc.*, **128**, 7720–7721.

25. Stankovich, S., Piner, R.D., Nguyen, S.T., and Ruoff, R.S. (2006) Synthesis and exfoliation of isocyanate-treated graphene oxide nanoplatelets. *Carbon*, **44**, 3342–3347.

26. Lomeda, J.R., Doyle, C.D., Kosynkin, D.V., Hwang, W.F., and Tour, J.M. (2008) Diazonium functionalization of surfactant-wrapped chemically converted graphene sheets. *J. Am. Chem. Soc.*, **130**, 16201–16206.

27. Shen, J., Hu, Y., Li, C., Qin, C., and Ye, M. (2009) Synthesis of amphiphilic graphene nanoplatelets. *Small*, **5**, 82–85.

28. Xu, Y., Bai, H., Lu, G., Li, C., and Shi, G. (2008) Flexible graphene films via the filtration of water-soluble noncovalent functionalized graphene sheets. *J. Am. Chem. Soc.*, **130**, 5856–5857.

29. Hao, R., Qian, W., Zhang, L., and Hou, Y. (2008) Aqueous dispersions of TCNQ anion stabilized graphene sheets. *Chem. Commun.*, 6576–6578.

30. Terrones, M., Botello-Méndez, A.R., Campos-Delgado, J., López-Urías, F., Vega-Cantú, Y.I., Rodríguez-Macías, F.J., Elías, A.L., Munoz-Sandoval, E., Cano-Márquez, A.G. *et al.* (2010) Graphene and graphite nanoribbons: morphology, properties, synthesis, defects and applications. *Nano Today*, **5**, 351–372.

31. Hwang, E.H., Adam, S., and Das Sarma, S. (2007) Carrier transport in two-dimensional graphene layers. *Phys. Rev. Lett.*, **98** (18), 6806–6806.

32. Novoselov, K.S., Morozov, S.V., Mohinddin, T.M.G., Ponomarenko, L.A., Elias, D.C., Yang, R., Barbolina, I.I., Blake, P., Booth, T.J., Jiang, D., Giesbers, J., Hill, E.W., and Geim, A.K. (2007) Electronic properties of graphene. *Phys. Status Solidi B: Basic Solid State Phys.*, **244**, 4106–4111.

33. Morozov, S.V., Novoselov, K.S., and Geim, A.K. (2008) Electron transport in graphene. *Phys. Usp.*, **51**, 744–748.

34. Bolotin, K.I., Sikes, K.J., Jiang, Z., Klima, M., Fudenberg, G., Hone, J., Kim, P., and Stormer, H.L. (2008) Ultrahigh electron mobility in suspended graphene. *Solid State Commun.*, **146**, 351–355.

35. Zhou, M., Zhai, Y.M., and Dong, S.J. (2009) Electrochemical sensing and biosensing platform based on chemically reduced graphene oxide. *Anal. Chem.*, **81**, 5603–5613.

36. McCreery, R.L. (2008) Advanced carbon electrode materials for molecular electrochemistry. *Chem. Rev.*, **108**, 2646–2687.

37. Niwa, O., Jia, J., Sato, Y., Kato, D., Kurita, R., Maruyama, K., Suzuki, K., and Hirono, S. (2006) Electrochemical performance of Angstrom level flat sputtered carbon film consisting of sp(2) and sp(3) mixed bonds. *J. Am. Chem. Soc.*, **128**, 7144–7145.

38. Yang, S.L., Guo, D.Y., Su, L., Yu, P., Li, D., Ye, J.S., and Mao, L.Q. (2009) A facile method for preparation of graphene film electrodes with tailor-made dimensions with Vaseline as the insulating binder. *Electrochem. Commun.*, **11**, 1912–1915.

39. Lin, W.J., Liao, C.S., Jhang, J.H., and Tsai, Y.C. (2009) Graphene modified basal and edge plane pyrolytic graphite electrodes for electrocatalytic oxidation of hydrogen peroxide and beta-nicotinamide adenine dinucleotide. *Electrochem. Commun.*, **11**, 2153–2156.

40. Tang, L.H., Wang, Y., Li, Y.M., Feng, H.B., Lu, J., and Li, J.H. (2009) Preparation, structure, and electrochemical properties of reduced graphene sheet films. *Adv. Funct. Mater.*, **19**, 2782–2789.

41. Shang, N.G., Papakonstantinou, P., McMullan, M., Chu, M., Stamboulis, A., Potenza, A., Dhesi, S.S., and Marchetto, H. (2008) Catalyst-free efficient growth, orientation and biosensing properties of multilayer graphene nanoflake films with sharp edge planes. *Adv. Funct. Mater.*, **18**, 3506–3514.

42. Wang, J.F., Yang, S.L., Guo, D.Y., Yu, P., Li, D., Ye, J.S., and Mao, L.Q. (2009) Comparative studies on electrochemical activity of

graphene nanosheets and carbon nanotubes. *Electrochem. Commun.*, **11**, 1892–1895.

43. Nicholson, R.S. (1965) Theory and application of cyclic voltammetry for measurement of electrode. Reaction kinetics. *Anal. Chem.*, **37**, 1351–1355.

44. Fischer, A.E., Show, Y., and Swain, G.M. (2004) Electrochemical performance of diamond thin-film electrodes from different commercial sources. *Anal. Chem.*, **76**, 2553–2560.

45. Yang, W., Ratinac, K.R., Ringer, S.P., Thordarson, P., Gooding, J.J., and Braet, F. (2010) Carbon nanomaterials in biosensors: should you use nanotubes or graphene? *Angew. Chem. Int. Ed. Engl.*, **49**, 2114–2138.

46. Alwarappan, S., Erdem, A., Liu, C., and Li, C.Z. (2009) Probing the electrochemical properties of graphene nanosheets for biosensing applications. *J. Biochem. Mol. Biol. Biophys.*, **113**, 8853–8857.

47. Banks, C.E., Davies, T.J., Wildgoose, G.G., and Compton, R.G. (2005) Electrocatalysis at graphite and carbon nanotube modified electrodes: edge-plane sites and tube ends are the reactive sites. *Chem. Commun.*, 829–841.

48. Chou, A., Bocking, T., Singh, N.K., and Gooding, J.J. (2005) Demonstration of the importance of oxygenated species at the ends of carbon nanotubes for their favourable electrochemical properties. *Chem. Commun.*, 842–844.

49. Wu, K.B., Fei, J.J., and Hu, S.S. (2003) Simultaneous determination of dopamine and serotonin on a glassy carbon electrode coated with a film of carbon nanotubes. *Anal. Biochem.*, **318**, 100–106.

50. Zhang, M.N., Gong, K.P., Zhang, H.W., and Mao, L.Q. (2005) Layer-by-layer assembled carbon nanotubes for selective determination of dopamine in the presence of ascorbic acid. *Biosens. Bioelectron.*, **20**, 1270–1276.

51. Yao, Y.L. and Shiu, K.K. (2008) Direct electrochemistry of glucose oxidase at carbon nanotube-gold colloid modified electrode with poly(diallyldimethylammonium chloride) coating. *Electroanalysis*, **20**, 1542–1548.

52. Leger, C. and Bertrand, P. (2008) Direct electrochemistry of redox enzymes as a tool for mechanistic studies. *Chem. Rev.*, **108**, 2379–2438.

53. Armstrong, F.A., Hill, H.A.O., and Walton, N.J. (1988) Direct electrochemistry of redox proteins. *Acc. Chem. Res.*, **21**, 407–413.

54. Zhang, W.J. and Li, G.X. (2004) Third-generation biosensors based on the direct electron transfer of proteins. *Anal. Sci.*, **20**, 603–609.

55. Wu, Y.H. and Hu, S.S. (2007) Biosensors based on direct electron transfer in redox proteins. *Microchim. Acta*, **159**, 1–17.

56. Prakash, P.A., Yogeswaran, U., and Chen, S.M. (2009) Direct electrochemistry of catalase at multiwalled carbon nanotubes-nafion in presence of needle shaped DDAB for H_2O_2 sensor. *Talanta*, **78**, 1414–1421.

57. Shan, C.S., Yang, H.F., Song, J.F., Han, D.X., Ivaska, A., and Niu, L. (2009) Direct electrochemistry of glucose oxidase and biosensing for glucose based on graphene. *Anal. Chem.*, **81**, 2378–2382.

58. Kang, X.H., Wang, J., Wu, H., Aksay, A.I., Liu, J., and Lin, Y.H. (2009) Glucose Oxidase-graphene-chitosan modified electrode for direct electrochemistry and glucose sensing. *Biosens. Bioelectron.*, **25**, 901–905.

59. Schniepp, H.C., Li, J.L., McAllister, M.J., Sai, H., Herrera-Alonso, M., Adamson, D.H., Prud_homme, R.K., Car, R., Saville, D.A., and Aksay, I.A. (2006) Functionalized single graphene sheets derived from splitting graphite oxide. *J. Biochem. Mol. Biol. Biophys.*, **110**, 8535–8539.

60. Geng, Y., Wang, S.J., and Kim, J.K. (2009) Preparation of graphite nanoplatelets and graphene sheets. *J. Colloid Interface Sci.*, **336**, 592–598.

61. Bekyarova, E., Itkis, M.E., Ramesh, P., Berger, C., Sprinkle, M., and Herr, W.A. (2009) Chemical modification of epitaxial graphene: spontaneous grafting of aryl groups. *J. Am. Chem. Soc.*, **131**, 1336–1337.

62. Shan, C., Yang, H., Han, D., Zhang, Q., Ivaska, A., and Niu, L. (2009) Water-soluble graphene covalently functionalized by biocompatible poly-L-lysine. *Langmuir*, **25**, 12030–12033.

63. Bai, H., Xu, Y., Zhao, L., Li, C., and Shi, G. (2009) Non-covalent functionalization of graphene sheets by sulfonated polyaniline. *Chem. Commun.*, 1667–1669.

64. Stankovich, S., Piner, R.D., Chen, X., Wu, N., Nguyen, S.T., and Ruoff, R.S. (2006) Stable aqueous dispersions of graphitic nanoplatelets via the reduction of exfoliated graphite oxide in the presence of poly(sodium 4-styrenesulfonate). *J. Mater. Chem.*, **16**, 155–158.

65. Salavagione, H.J., Gomez, M.A., and Martinez, G. (2009) Polymeric modification of graphene through esterification of graphite oxide and poly(vinyl alcohol). *Macromolecules*, **42**, 6331–6334.

66. Worsley, K.A., Ramesh, P., Mandal, S.K., Niyogi, S., Itkis, M.E., and Haddon, R.C. (2007) Soluble graphene derived from graphite fluoride. *Chem. Phys. Lett.*, **445**, 51–56.

67. Liu, N., Luo, F., Wu, H., Liu, Y., Zhang, C., and Chen, J. (2008) One-step ionic-liquid-assisted electrochemical synthesis of ionic-liquid-functionalized graphene sheets directly from graphite. *Adv. Funct. Mater.*, **18**, 1518–1525.

68. Bon, S.B., Valentini, L., Verdejo, R., Garcia Fierro, J.L., Peponi, L., Lopez-Manchado, M.A., and Kenny, J.M. (2009) Plasma fluorination of chemically derived graphene sheets and subsequent modification with butylamine. *Chem. Mater.*, **21**, 3433–3438.

69. Osuna, S., Torrent-Sucarrat, M., Solà, M., Geerlings, P., Ewels, C.P., and Lier, G.V. (2010) Reaction mechanisms for graphene and carbon nanotube fluorination. *J. Biochem. Mol. Biol. Biophys.*, **114**, 3340–3345.

70. Wang, B., Cooley, B.J., Cheng, S., Zou, K., Hao, Q.Z., Okino, F., Sofo, J., Samarth, N., and Zhu, J. (2010) Opening a bandgap in graphene by fluorination APS March Meeting 2010 Abstracts, American Physical Society.

71. Sofo, J.O., Chaudhari, A.S., and Barber, G.D. (2007) Graphane: A two-dimensional hydrocarbon. *Phys. Rev. B*, **75**, 3401–3401.

72. Elias, D.C., Nair, R.R., Mohiuddin, T.M.G., Morozov, S.V., Blake, P., Halsall, M.P., Ferrari, A.C., Boukhvalov, D.W., and Katsnelson, M.I. (2009) Control of graphene's properties by reversible hydrogenation: evidence for graphane. *Science*, **323**, 610–613.

73. Sessi, P., Guest, J.R., Bode, M., and Guisinger, N.P. (2009) Patterning graphene at the nanometer scale via hydrogen desorption. *Nano Lett.*, **9**, 4343–4347.

74. Lee, C., Wei, X., Kysar, J.W., and Hone, J. (2008) Measurement of the elastic properties and intrinsic strength of monolayer graphene. *Science*, **321**, 385–388.

75. Ferralis, N. (2010) Probing mechanical properties of graphene with Raman spectroscopy. *J. Mater. Sci.*, **45**, 5135–5149.

76. Frank, I., Tanenbaum, D., Van der Zande, A., and McEuen, P. (2007) Mechanical properties of suspended graphene sheets. *J. Vac. Sci. Technol. B*, **25**, 2558–2561.

77. Novoselov, K.S., Jiang, D., Schedin, F., Booth, T., Khotkevich, V., Morozov, S., and Geim, A.K. (2005) Two-dimensional atomic crystals.

Proc. Natl. Acad. Sci. U.S.A., **102**, 10451–10453.

78. Calizo, I., Bao, W., Miao, F., Lau, C., and Balandin, A. (2007) The effect of substrates on the Raman spectrum of graphene: graphene-on-sapphire and graphene-on-glass. *Appl. Phys. Lett.*, **91**, 1904–1904.

79. Kim, K., Zhao, Y., Jang, H., Lee, S., Kim, J., Kim, K., Ahn, J.H., Kim, P., Choi, J.Y., and Hong, B. (2009) Large-scale pattern growth of graphene films for stretchable transparent electrodes. *Nature*, **457**, 706–710.

80. Mohiuddin, T., Lombardo, A., Nair, R., Bonetti, A., Savini, G., Jalil, R., Bonini, N., Basko, D., Galiotis, C., Marzari, N., Novoselov, K., Geim, A.K., and Ferrari, A. (2009) Uniaxial strain in graphene by Raman spectroscopy: G peak splitting, Gruneisen parameters, and sample orientation. *Phys. Rev. B*, **79**, 5433–5433.

81. Hass, J., de Heer, W., and Conrad, E. (2008) The growth and morphology of epitaxial multilayer graphene. *J. Phys. Condens. Matter.*, **20**, 23202–23202.

82. Berger, C., Song, Z., Li, T., Li, X., Ogbazghi, A., Feng, R., Dai, Z., Marchenkov, A., Conrad, E., First, P., and de Heer, W. (2004) Ultrathin epitaxial graphite: 2D electron gas properties and a route toward graphene-based nanoelectronics. *J. Biochem. Mol. Biol. Biophys.*, **108**, 19912–19916.

83. Marchini, S., Gunther, S., and Wintterlin, J. (2007) Scanning tunneling microscopy of graphene on Ru(0001). *Phys. Rev. B*, **76**, 5429–5429.

84. de Parga, V.A., Calleja, F., Borca, B., Passeggi, M., Hinarejos, J., Guinea, F., and Miranda, R. (2008) Periodically rippled graphene: growth and spatially resolved electronic structure. *Phys. Rev. Lett.*, **100**, 6807–6807.

85. Ferralis, N., Maboudian, R., and Carraro, C. (2008) Evidence of structural strain in epitaxial graphene layers on 6H-SiC(0001). *Phys. Rev. Lett.*, **101**, 6801–6801.

86. Calizo, I., Balandin, A.A., Bao, W., Miao, F., and Lau, C. (2007) Temperature dependence of the Raman spectra of graphene and graphene multilayers. *Nano Lett.*, **7**, 2645–2649.

87. Calizo, I., Miao, F., Bao, W., Lau, C., and Balandin, A. (2007) Variable temperature Raman microscopy as a nanometrology tool for graphene layers and graphene-based devices. *Appl. Phys. Lett.*, **91**, 71913–71913.

88. Yu, M., Lourie, O., Dyer, M.J., Kelly, T.F., and Ruoff, R.S. (2000) Strength and breaking mechanism of multiwalled carbon nanotubes under tensile load. *Science*, **287**, 637–640.

89. Frank, S.P., Poncharal, P., Wang, Z.L., and de Heer, W.A. (1998) Carbon nanotube quantum resistors. *Science*, **280**, 1744–1746.

90. Pandolfo, A.G. and Hollenkamp, A.F. (2006) Carbon properties and their role in supercapacitors. *J. Power Sources*, **157**, 11–27.

91. Noel, M. and Suryanarayanan, V. (2002) Role of carbon host lattices in Li-ion intercalation/de-intercalation processes? *J. Power Sources*, **111**, 193–209.

92. Dicks, A.L. (2006) The role of carbon in fuel cells. *J. Power Sources*, **156**, 128–141.

93. Baughman, R.H., Cui, C., Zakhidov, A.A., Iqbal, Z., Barisci, J.N., Spinks, G.M., Wallace, G.G., Mazzoldi, A., De Rossi, D., Rinzler, A.G., Jaschinski, O., Roth, S., and Kertesz, M. (1999) Carbon nanotube actuators. *Science*, **284**, 1340–1344.

94. Iijima, S. (1991) Helical microtubules of graphitic carbon. *Nature*, **354**, 56–58.

95. Hou, P., Liu, C., Tong, Y., Xu, S., Liu, M., and Cheng, H. (2001) Purification of single-walled carbon nanotubes synthesized by the

hydrogen arc-discharge method. *J. Mater. Res.*, **16**, 2526–2529.

96. Maser, W.K., Muñoz, E., Benito, A.M., Martínez, M.T., de la Fuente, G.F., Maniette, Y., Anglaret, E., and Sauvajol, J.L. (1998) Production of high-density single-walled nanotube material by a simple laser-ablation method. *Chem. Phys. Lett.*, **292**, 587–593.

97. Baker, R.T.K. and Rodriguez, N.M. (1994) Catalytic growth of carbon nanofibers and nanotubes. Symposium of the MRS, Materials Research Society.

98. Shen, C., Brozena, A.H., and Wang, Y.H. (2001) Double-walled carbon nanotubes: challenges and opportunities. *Nanoscale*, **3**, 503–518.

99. Journet, C., Matser, W.K., Bernier, P., Laiseau, L., Lefrant, S., Deniard, P., Lee, R., and Fischer, J.E. (1997) Large-scale production of single-walled carbon nanotubes by the electric-arc technique. *Nature*, **388**, 756–758.

100. Thess, A., Lee, R., Nikolaev, P., Dai, H., Petit, P., Robert, J., Xu, C., Lee, Y.H., and Smalley, R.E. (1996) Crystalline ropes of metallic carbon nanotubes. *Science*, **276**, 483–487.

101. Chen, J., Liu, Y., Minett, A.I., Lynam, C., Wang, J., and Wallace, G.G. (2007) Flexible, aligned carbon nanotube/conducting polymer electrodes for a lithium-ion battery. *Chem. Mater.*, **19** (15), 3595–3597.

102. Duesberg, G.S., Muster, J., Krstic, V., Burghard, M., and Roth, S. (1998) Chromatographic size separation of single-wall carbon nanotubes. *Appl. Phys. A. Mater. Sci. Proc.*, **67**, 117–119.

103. Duesberg, G.S., Blau, W., Byrne, H.J., Muster, J., Burghard, M., and Roth, S. (1999) Chromatography of carbon nanotubes. *Synth. Met.*, **103**, 2484–2485.

104. Arnold, M.S., Green, A.A., Hulvat, J.F., Stupp, S.I., and Hersam, M.C. (2006) Sorting carbon nanotubes by electronic structure using density

105. Lynam, C., Minett, A.I., Habas, S.E., Gambhir, S., Officer, D.L., and Wallace, G.G. (2008) Functionalising carbon nanotubes. *Int. J. Nanotech.*, **5** (2), 331–351.

106. Islam, M.F., Rojas, E., Bergey, D.M., Johnson, A.T., and Yodh, A.G. (2003) High weight fraction surfactant solubilization of single-wall carbon nanotubes in water. *Nano Lett.*, **3**, 269–273.

107. Vigolo, B., Penicaud, A., Coulon, C., Sauder, C., Pailler, R., Jounet, C., Bernier, P., and Poulin, P. (2000) Macroscopic fibers and ribbons of oriented carbon nanotubes. *Science*, **290**, 1331–1334.

108. Regev, O., ElKati, P.N.B., Loos, J., and Koning, C.E. (2004) Preparation of conductive nanotube-polymer composites using latex technology. *Adv. Mater.*, **16**, 248–251.

109. Avouris, P., Appenzeller, J., Martel, R., and Wind, S.J. (2003) Carbon nanotube electronics. *Proc. IEEE*, **91**, 1772–1784.

110. Avouris, P. (2002) Molecular electronics with carbon nanotubes. *Acc. Chem. Res.*, **35**, 1026–1034.

111. Di Carlo, A., Pecchia, A., Petrolati, E., and Paoloni, C. (2006) Modelling of carbon nanotube-based devices: from nanoFETs to THz emitters. *Nanomodeling II*, **6328**, 32808–32808 (Art. No.: 632808).

112. McEuen, P.L. and Park, J.Y. (2004) Electron transport in single-walled carbon nanotubes. *MRS Bull.*, **29**, 272–275.

113. Imry, Y. and Landauer, R. (1999) Conductance viewed as transmission. *Rev. Mod. Phys.*, **71**, S306–S312.

114. McEuen, P.L., Bockrath, M., Cobden, D.H., Yoon, Y.G., and Louie, S.G. (1999) Disorder, pseudospins, and backscattering in carbon nanotubes. *Phys. Rev. Lett.*, **83**, 5098–5101.

115. Kong, J., Yenilmez, E., Tombler, T.W., Kim, W., Dai, H.J., Laughlin,

R.B., Liu, L., Jayanthi, C.S., and Wu, S.Y. (2001) Quantum interference and ballistic transmission in nanotube electron waveguides. *Phys. Rev. Lett.* **87**, 106801–106804, (Art. No.: 106801).

116. Shiraishi, M. and Ata, M. (2001) Work function of carbon nanotubes. *Carbon*, **39**, 1913–1917.

117. Mann, D., Javey, A., Kong, J., Wang, Q., and Dai, H.J. (2003) Ballistic transport in metallic nanotubes with reliable Pd ohmic contacts. *Nano Lett.*, **3**, 1541–1544.

118. Stahl, H., Appenzeller, J., Martel, R., Avouris, P., and Lengeler, B. (2000) Intertube coupling in ropes of single-wall carbon nanotubes. *Phys. Rev. Lett.*, **85**, 5186–5189.

119. Kong, J., Zhou, C., Morpurgo, A., Soh, H.T., Quate, C.F., Marcus, C., and Dai, H. (1999) Synthesis, integration, and electrical properties of individual single-walled carbon nanotubes. *Appl. Phys. A-Mater. Sci. Process*, **69**, 305–308.

120. Zhou, C.W., Kong, J., and Dai, H.J. (2000) Electrical measurements of individual semiconducting single-walled carbon nanotubes of various diameters. *Appl. Phys. Lett.*, **76**, 1597–1599.

121. Zhou, C.W., Kong, J., and Dai, H.J. (2000) Intrinsic electrical properties of individual single-walled carbon nanotubes with small band gaps. *Phys. Rev. Lett.*, **84**, 5604–5607.

122. Soh, H.T., Quate, C.F., Morpurgo, A.F., Marcus, C.M., Kong, J., and Dai, H.J. (1999) Integrated nanotube circuits: controlled growth and ohmic contacting of single-walled carbon nanotubes. *Appl. Phys. Lett.*, **75**, 627–629.

123. Nihei, M., Kawabata, A., Kondo, D., Horibe, M., Sato, S., and Awano, Y. (2005) Electrical properties of carbon nanotube bundles for future via interconnects. *Jpn. J. App. Phys. Part 1-Reg. Pap. Short Notes Rev. Pap.*, **44**, 1626–1628.

124. Naeemi, A. and Meindl, J.D. (2007) Design and performance modeling for single-walled carbon nanotubes as local, semiglobal, and global interconnects in gigascale integrated systems. *IEEE Trans. Electron Devices*, **54**, 26–37.

125. Talapatra, S., Kar, S., Pal, S.K., Vajtai, R., Ci, L., Victor, P., Shaijumon, M.M., Kaur, S., Nalamasu, O., and Ajayan, P.M. (2006) Direct growth of aligned carbon nanotubes on bulk metals. *Nat. Nanotech.*, **1**, 112–116.

126. Clifford, J.P., John, D.L., Castro, L.C., and Pulfrey, D.L. (2004) Electrostatics of partially gated carbon nanotube FETs. *IEEE Trans. Nanotech.*, **3**, 281–286.

127. John, D.L., Castro, L.C., Clifford, J., and Pulfrey, D.L. (2003) Electrostatics of coaxial Schottky-barrier nanotube field-effect transistors. *IEEE Trans. Nanotech.*, **2**, 175–180.

128. Kajiura, H., Nandyala, A., and Bezryadin, A. (2005) Quasi-ballistic electron transport in as-produced and annealed multiwall carbon nanotubes. *Carbon*, **43**, 1317–1319.

129. Nessim, G.D., Seita, M., O'Brien, K.P., Hart, A.J., Bonaparte, R.K., Mitchell, R.R., and Thompson, C.V. (2009) Low temperature synthesis of vertically aligned carbon nanotubes with electrical contact to metallic substrates enabled by thermal decomposition of the carbon feedstock. *Nano Lett.*, **9**, 3398–3405.

130. Collins, P.C., Arnold, M.S., and Avouris, P. (2001) Engineering carbon nanotubes and nanotube circuits using electrical breakdown. *Science*, **292**, 706–709.

131. Collins, P.G. and Avouris, P. (2002) Multishell conduction in multiwalled carbon nanotubes. *App. Phys. A-Mater. Sci. Process*, **74**, 329–332.

132. Bachtold, A., Strunk, C., Salvetat, J.P., Bonard, J.M., Forro, L., Nussbaumer, T., and Schonenberger, C. (1999) Aharonov-Bohm oscillations in carbon nanotubes. *Nature*, **397**, 673–675.

133. Schonenberger, C., Bachtold, A., Strunk, C., Salvetat, J.P., and Forro, L. (1999) Interference and Interaction in multi-wall carbon nanotubes. *Appl. Phys. A-Mater. Sci. Process*, **69**, 283–295.

134. Li, H.J., Lu, W.G., Li, J.J., Bai, X.D., and Gu, C.Z. (2005) Multichannel ballistic transport in multiwall carbon nanotubes. *Phys. Rev. Lett.*, **95**, 086601.

135. Moore, R.R., Banks, C.E., and Compton, R.G. (2004) Basal plane pyrolytic graphite modified electrodes: comparison of carbon nanotubes and graphite powder as electrocatalysts. *Anal. Chem.*, **76**, 2677–2682.

136. Banks, C.E., Moore, R.R., Davies, T.J., and Compton, R.G. (2004) Investigation of modified basal plane pyrolytic graphite electrodes: definitive evidence for the electrocatalytic properties of the ends of carbon nanotubes. *Chem. Commun.*, 1804–1805.

137. Davies, T.J., Hyde, M.E., and Compton, R.G. (2005) Nanotrench arrays reveal insight into graphite electrochemistry. *Angew. Chem. Int. Ed. Engl.*, **44**, 5121–5126.

138. Holloway, A.F., Wildgoose, G.G., Compton, R.G., Shao, L.D., and Green, M.L.H. (2008) The influence of edge-plane defects and oxygen-containing surface groups on the voltammetry of acid-treated, annealed and ''super-annealed'' multiwalled carbon nanotubes. *J. Solid State Electrochem.*, **12**, 1337–1348.

139. Pumera, M., Sasaki, T., and Iwai, H. (2008) Relationship between carbon nanotube structure and electrochemical behavior: heterogeneous electron transfer at electrochemically activated carbon nanotubes. *Chem. Asian J.*, **3**, 2046–2055.

140. Sanchez, S., Fabregas, E., and Pumera, M. (2009) Electrochemical activation of carbon nanotube/polymer composites. *Phys. Chem. Chem. Phys.*, **11**, 182–186.

141. Holloway, A.F., Toghill, K., Wildgoose, G.G., Compton, R.G., Ward, M.A.H., Tobias, G., Llewellyn, S.A., Ballesteros, B., Green, M.L.H., and Crossley, A. (2008) Electrochemical opening of single-walled carbon nanotubes filled with metal halides and with closed ends. *J. Biochem. Mol. Biol. Biophys.*, **112**, 10389–10397.

142. Musameh, M., Lawrence, N.S., and Wang, J. (2005) Electrochemical activation of carbon nanotubes. *Electrochem. Commun.*, **7**, 14–18.

143. Gong, K.P., Chakrabarti, S., and Dai, L.M. (2008) Electrochemistry at carbon nanotube electrodes: is the nanotube tip more active than the sidewall? *Angew. Chem. Int. Ed. Engl.*, **47**, 5446–5450.

144. Kolosnjaj, J., Szwarc, H., and Moussa, F. (2007) Toxicity studies of carbon nanotubes. *Bio-App. Nanopart.*, **620**, 181–204.

145. Lam, C.W., James, J.T., McCluskey, R., and Hunter, R.L. (2004) Pulmonary toxicity of single-wall carbon nanotubes in mice 7 and 90 days after intratracheal installation. *Toxicol. Sci.*, **77**, 126–134.

146. Qu, G.B., Bai, Y.H., Zhang, Y., Jia, Q., Zhang, W.D., and Yan, B. (2009) The effect of multiwalled carbon nanotube agglomeration on their accumulation in and damage to organs in mice. *Carbon*, **47**, 2060–2069.

147. Wang, X., Zang, J.J., Wang, H., Nie, H., Wang, T.C., Deng, X.Y., Gu, Y.Q., Liu, Z.H., and Jia, G. (2010) Pulmonary toxicity in mice exposed to low and medium doses of water-soluble multi-walled carbon nanotubes. *J. Nanosci. Nanotechnol.*, **10**, 8516–8526.

148. Ghodake, G., Seo, Y.D., Park, D., and Lee, D.S. (2010) Phytotoxicity of carbon nanotubes assessed by Brassica juncea and Phaseolus mungo. *J. Nanoelectron. Optoelectron.* **5**, 157–160.

149. Goyal, D., Zhang, X.J., and Rooney-Varga, J.N. (2010) Impacts of single-walled carbon nanotubes on microbial community structure in activated sludge. *Lett. Appl. Microbiol.* **51**, 428–435.

150. Shvedova, A.A., Castranova, V., Kisin, E.R., Schwegler-Berry, D., Murray, A.R., Gandelsman, V.Z., Maynard, A., and Baron, P. (2003) Exposure to carbon nanotube material: assessment of nanotube cytotoxicity using human keratinocyte cells. *J. Toxicol. Environ. Health-Part A*, **66**, 1909–1926.

151. Cherukuri, P., Bachilo, S.M., Litovsky, S.H., and Weisman, R.B. (2004) Near-infrared fluorescence microscopy of single-walled carbon nanotubes in phagocytic cells. *J. Am. Chem. Soc.*, **126**, 15638–15639.

152. Cui, D.X., Tian, F.R., Ozkan, C.S., Wang, M., and Gao, H.J. (2005) Effect of single wall carbon nanotubes on human HEK293 cells. *Toxicol. Lett.*, **155**, 73–85.

153. Kam, N.W.S., O'Connell, M., Wisdom, J.A., and Dai, H.J. (2005) Carbon nanotubes as multifunctional biological transporters and near-infrared agents for selective cancer cell destruction. *Proc. Nat. Acad. Sci. U.S.A.*, **102**, 11600–11605.

154. Sayes, C.M., Liang, F., Hudson, J.L., Mendez, J., Guo, W.H., Beach, J.M., Moore, V.C., Doyle, C.D., West, J.L., Billups, W.E. *et al.* (2006) Functionalization density dependence of single-walled carbon nanotubes cytotoxicity in vitro. *Toxicol. Lett.*, **161**, 135–142.

155. Zhao, Y.L., Xing, G.M., and Chai, Z.F. (2008) Nanotoxicology: are carbon nanotubes safe? *Nat. Nanotech.*, **3**, 191–192.

156. Barisci, J.N., Tahhan, M., Wallace, G.G., Badaire, S., Vaugien, T., Maugey, M., and Poulin, P. (2004) Properties of carbon nanotube fibers spun from DNA-stabilized dispersions. *Adv. Func. Mater.*, **14**, 133–138.

157. Dieckmann, G.R., Dalton, A.B., Johnson, P.A., Razal, J., Chen, J., Giordano, G.M., Munoz, E., Musselman, I.H., Baughman, R.H., and Draper, R.K. (2003) Controlled assembly of carbon nanotubes by designed amphiphilic peptide helices. *J. Am. Chem. Soc.*, **125**, 1770–1777.

158. Dalton, A.B., Ortiz-Acevedo, A., Zorbas, V., Brunner, E., Sampson, W.M., Collins, L., Razal, J.M., Yoshida, M.M., Baughman, R.H., Draper, R.K. *et al.* (2004) Hierarchical self-assembly of peptide-coated carbon nanotubes. *Adv. Funct. Mater.*, **14**, 1147–1151.

159. McCarthy, B., Dalton, A.B., Coleman, J.N., Byrne, H.J., Bernier, P., and Blau, W.J. (2001) Spectroscopic investigation of conjugated polymer/single walled carbon nanotube interactions. *Chem. Phys. Lett.*, **350**, 27–32.

160. Curran, S.A., Ajayan, P.M., Blau, W.J., Carroll, D.L., Coleman, J.N., Dalton, A.B., Davey, A.P., Drury, A., McCarthy, B., Maier, S., and Strevens, A. (1998) A composite from poly(m-phenylenevinylene-co-2,5-dioctoxy-p-phenylenevinylene) and carbon nanotubes: a novel material for molecular optoelectronics. *Adv. Mater.*, **10**, 1091–1093.

161. Mawhinney, D.B., Naumenko, V., Kuznetsova, A., Yates, J.T., Liu, J., and Smalley, R.E. (2000) Surface defect site density on single walled carbon nanotubes by titration. *Chem. Phys. Lett.*, **324**, 213–216.

162. Hu, H., Bhowmik, P., Zhao, B., Hamon, M.A., Itkis, M.E., and Haddon, R.C. (2001) Determination of the acidic sites of purified single-walled carbon nanotubes by acid-base titration. *Chem. Phys. Lett.*, **345**, 25–28.

163. Ramanathan, T., Fisher, F.T., Ruoff, R.S., and Brinson, L.C. (2005) Amino-functionalized carbon nanotubes for binding to polymers and biological systems. *Chem. Mater.*, **17**, 1290–1295.

164. Hamon, M.A., Hui, H., Bhowmik, P., Itkis, H.M.E., and Haddon, R.C. (2002) Ester-functionalized soluble single-walled carbon nanotubes. *App. Phys. A-Mater. Sci. Process*, **74**, 333–338.

165. Riggs, J.E., Walker, D.B., Carroll, D.L., and Sun, Y.P. (2000) Optical limiting properties of suspended and solubilized carbon nanotubes. *J. Biochem. Mol. Biol. Biophys.*, **104**, 7071–7076.

166. Chiu, P.W., Duesberg, G.S., Dettlaff-Weglikowska, U., and Roth, S. (2002) Interconnection of carbon nanotubes by chemical functionalization. *Appl. Phys. Lett.*, **80**, 3811–3813.

167. Frehill, F., Vos, J.G., Benrezzak, S., Koos, A.A., Konya, Z., Ruther, M.G., Blau, W.J., Fonseca, A., Nagy, J.B., Biro, L.P. *et al.* (2002) Interconnecting carbon nanotubes with an inorganic metal complex. *J. Am. Chem. Soc.*, **124**, 13694–13695.

168. Lim, J.K., Yun, W.S., Yoon, M.H., Lee, S.K., Kim, C.H., Kim, K., and Kim, S.K. (2003) Selective thiolation of single-walled carbon nanotubes. *Synth. Met.*, **139**, 521–527.

169. Fu, K.F., Huang, W.J., Lin, Y., Zhang, D.H., Hanks, T.W., Rao, A.M., and Sun, Y.P. (2002) Functionalization of carbon nanotubes with bovine serum albumin in homogeneous aqueous solution. *J. Nanosci. Nanotechnol.*, **2**, 457–461.

170. Huang, W.J., Taylor, S., Fu, K.F., Lin, Y., Zhang, D.H., Hanks, T.W., Rao, A.M., and Sun, Y.P. (2002) Attaching proteins to carbon nanotubes via diimide-activated amidation. *Nano Lett.*, **2**, 311–314.

171. Williams, K.A., Veenhuizen, P.T.M., de la Torre, B.G., Eritja, R., and Dekker, C. (2002) Nanotechnology – carbon nanotubes with DNA recognition. *Nature*, **420**, 761–761.

172. Dwyer, C., Guthold, M., Falvo, M., Washburn, S., Superfine, R., and Erie, D. (2002) DNA-functionalized single-walled carbon nanotubes. *Nanotechnology*, **13**, 601–604.

173. Lee, C.S., Baker, S.E., Marcus, M.S., Yang, W.S., Eriksson, M.A., and Hamers, R.J. (2004) Electrically addressable biomolecular functionalization of carbon nanotube and carbon nanofiber electrodes. *Nano Lett.*, **4**, 1713–1716.

174. Liu, J.Q., Chou, A., Rahmat, W., Paddon-Row, M.N., and Gooding, J.J. (2005) Achieving direct electrical connection to glucose oxidase using aligned single walled carbon nanotube arrays. *Electroanalysis*, **17**, 38–46.

175. Pompeo, F. and Resasco, D.E. (2002) Water solubilization of single-walled carbon nanotubes by functionalization with glucosarnine? *Nano Lett.*, **2**, 369–373.

176. Boul, P.J., Liu, J., Mickelson, E.T., Huffman, C.B., Ericson, L.M., Chiang, I.W., Smith, K.A., Colbert, D.T., Hauge, R.H., Margrave, J.L., and Smalley, R.E. (1999) Reversible sidewall functionalization of buckytubes. *Chem. Phys. Lett.*, **310**, 367–372.

177. Khabashesku, V.N., Billups, W.E., and Margrave, J.L. (2002) Fluorination of single-wall carbon nanotubes and subsequent derivatization reactions. *Acc. Chem. Res.*, **35**, 1087–1095.

178. Holzinger, M., Vostrowsky, O., Hirsch, A., Hennrich, F., Kappes, M., Weiss, R., and Jellen, F. (2001) Sidewall functionalization of carbon nanotubes. *Angew. Chem. Int. Ed. Engl.*, **40**, 4002–4005.

179. Kooi, S.E., Schlecht, U., Burghard, M., and Kern, K. (2002) Electrochemical modification of single carbon nanotubes *Angew. Chem.-Int. Ed.*, **41**, 1353–1355.

180. Hu, H., Ni, Y.C., Montana, V., Haddon, R.C., and Parpura, V. (2004) Chemically functionalized carbon nanotubes as substrates for neuronal growth. *Nano Lett.*, **4**, 507–511.

181. Kong, H., Gao, C., and Yan, D.Y. (2004) Functionalization of multi-walled carbon nanotubes by atom transfer radical polymerization and defunctionalization of the products. *Macromolecules*, **37**, 4022–4030.

182. Liu, Y., Wu, D.C., Zhang, W.D., Jiang, X., He, C.B., Chung, T.S., Goh, S.H., and Leong, K.W. (2005) Polyethylenimine-grafted multiwalled carbon nanotubes for secure noncovalent immobilization and efficient delivery of DNA. *Angew. Chem. Int. Ed. Engl.*, **44**, 4782–4785.

183. Zhang, X.F., Liu, T., Sreekumar, T.V., Kumar, S., Moore, V.C., Hauge, R.H., and Smalley, R.E. (2003) Poly(vinyl alcohol)/SWNT composite film. *Nano Lett.*, **3**, 1285–1288.

184. In Het Panhuis, M., Salvador-Morales, C., Franklin, E., Chambers, G., Fonseca, A., Nagy, J.B., Blau, W.J., and Minett, A.I. (2003) Characterization of an interaction between functionalized carbon nanotubes and an enzyme. *J. Nanosci. Nanotechnol.*, **3**, 209–213.

185. O'Connell, M.J., Bachilo, S.M., Huffman, C.B., Moore, V.C., Strano, M.S., Haroz, E.H., Rialon, K.L., Boul, P.J., Noon, W.H., Kittrell, C. *et al.* (2002) Band gap fluorescence from individual single-walled carbon nanotubes. *Science*, **297**, 593–596.

186. Murakami, H., Nomura, T., and Nakashima, N. (2003) Noncovalent porphyrin-functionalized single-walled carbon nanotubes in solution and the formation of porphyrin-nanotube nanocomposites. *Chem. Phys. Lett.*, **378**, 481–485.

187. Star, A., Stoddart, J.F., Steuerman, D., Diehl, M., Boukai, A., Wong, E.W., Yang, X., Chung, S.W., Choi, H., and Heath, J.R. (2001) Preparation and properties of polymer-wrapped single-walled carbon nanotubes. *Angew. Chem. Int. Ed. Engl.*, **40**, 1721–1725.

188. Coleman, J.N., Dalton, A.B., Curran, S., Rubio, A., Davey, A.P., Drury, A., McCarthy, B., Lahr, B., Ajayan, P.M., Roth, S. *et al.* (2000) Phase separation of carbon nanotubes and turbostratic graphite using a functional organic polymer. *Adv. Mater.*, **12**, 213–217.

189. Moore, V.C., Strano, M.S., Haroz, E.H., Hauge, R.H., Smalley, R.E., Schmidt, J., and Talmon, Y. (2003) Individually suspended single-walled carbon nanotubes in various surfactants. *Nano Lett.*, **3**, 1379–1382.

190. Zhang, X.F., Liu, T., Sreekumar, T.V., Kumar, S., Moore, V.C., Hauge, R.H., and Smalley, R.E. (2003) Poly(vinyl alcohol)/SWNT Composite Film *Nano Lett.*, **3** (9), 1285–1288.

191. Wang, J., Musameh, M., and Lin, Y.H. (2003) Solubilization of carbon nanotubes by Nafion toward the preparation of amperometric biosensors. *J. Am. Chem. Soc.*, **125**, 2408–2409.

192. Arnold, M.S., Stupp, S.I., and Hersam, M.C. (2005) Enrichment of single-walled carbon nanotubes by diameter in density gradients. *Nano Lett.*, **5**, 713–718.

193. Guo, Z.J., Sadler, P.J., and Tsang, S.C. (1998) Immobilization and visualization of DNA and proteins on carbon nanotubes. *Adv. Mater.*, **10**, 701–703.

194. Chen, R.J., Zhang, Y.G., Wang, D.W., and Dai, H.J. (2001) Noncovalent sidewall functionalization of single-walled carbon nanotubes for protein immobilization. *J. Am. Chem. Soc.*, **123**, 3838–3839.

195. Balavoine, F., Schultz, P., Richard, C., Mallouh, V., Ebbesen, T.W., and Mioskowski, C. (1999) Helical crystallization of proteins on carbon nanotubes: a first step towards the development of new biosensors. *Angew. Chem. Int. Ed. Engl.*, **38**, 1912–1915.

196. Davis, J.J., Coleman, K.S., Azamian, B.R., Bagshaw, C.B.,

and Green, M.L.H. (2003) Chemical and biochemical sensing with modified single walled carbon nanotubes. *Chem.-A Eur. J.*, **9**, 3732–3739.

197. Wang, S.Q., Humphreys, E.S., Chung, S.Y., Delduco, D.F., Lustig, S.R., Wang, H., Parker, K.N., Rizzo, N.W., Subramoney, S., Chiang, Y.M. *et al.* (2003) Peptides with selective affinity for carbon nanotubes. *Nat. Mater.*, **2**, 196–200.

198. Kim, O.K., Je, J.T., Baldwin, J.W., Kooi, S., Pehrsson, P.E., and Buckley, L.J. (2003) Solubilization of single-wall carbon nanotubes by supramolecular encapsulation of helical amylose. *J. Am. Chem. Soc.*, **125**, 4426–4427.

199. Pantarotto, D., Partidos, C.D., Graff, R., Hoebeke, J., Briand, J.P., Prato, M., and Bianco, A. (2003) Synthesis, structural characterization, and immunological properties of carbon nanotubes functionalized with peptides. *J. Am. Chem. Soc.*, **125**, 6160–6164.

200. Britto, P.J., Santhanam, K.S.V., and Ajayan, P.M. (1996) Carbon nanotube electrode for oxidation of dopamine. *Bioelectrochem. Bioenerg.*, **41**, 121–125.

201. Britto, P.J., Santhanam, K.S.V., Rubio, A., Alonso, J.A., and Ajayan, P.M. (1999) Improved charge transfer at carbon nanotube electrodes. *Adv. Mater.*, **11**, 154–157.

202. Bandyopadhyaya, R., Nativ-Roth, E., Regev, O., and Yerushalmi-Rozen, R. (2002) Stabilization of individual carbon nanotubes in aqueous solutions. *Nano Lett.*, **2**, 25–28.

203. Supronowicz, P.R., Ajayan, P.M., Ullmann, K.R., Arulanandam, B.P., Metzger, D.W., and Bizios, R. (2002) Novel current-conducting composite substrates for exposing osteoblasts to alternating current stimulation. *J. Biomed. Mater. Res.*, **59**, 499–506.

204. Moulton, S.E., Minett, A.I., Murphy, R., Ryan, K.P., McCarthy, D., Coleman, J.N., Blau, W.J., and Wallace, G.G. (2005) Biomolecules as selective dispersants for carbon nanotubes. *Carbon*, **43** (9), 1879–1884.

205. Katz, E. and Willner, I. (2004) Biomolecule-functionalized carbon nanotubes: applications in nanobioelectronics. *Chem. Phys. Chem.*, **5**, 1085–1104.

206. Nayagam, D.A.X., Williams, R.A., Chen, J., Magee, K.A., Irwin, J., Tan, J., Innis, P., Leung, R.T., Finch, S., Williams, C.E. *et al.* (2011) Biocompatibility of immobilized aligned carbon nanotubes. *Small*, **7**, 1035–1042.

207. Balasubramanian, K. and Burghard, M. (2005) Chemically functionalized carbon nanotubes. *Small*, **1**, 180–192.

208. Pantarotto, D., Briand, J.P., Prato, M., and Bianco, A. (2004) Translocation of bioactive peptides across cell membranes by carbon nanotubes. *Chem. Commun.*, 16–17.

209. Kam, N.W.S., Liu, Z.A., and Dai, H.J. (2006) Carbon nanotubes as intracellular transporters for proteins and DNA: an investigation of the uptake mechanism and pathway. *Angew. Chem. Int. Ed. Engl.*, **45**, 577–581.

210. Lacerda, L., Raffa, S., Prato, M., Bianco, A., and Kostarelos, K. (2007) Cell-penetrating CNTs for delivery of therapeutics. *Nano Today*, **2**, 38–43.

211. Prato, M., Kostarelos, K., and Bianco, A. (2008) Functionalized carbon nanotubes in drug design and discovery. *Acc. Chem. Res.*, **41**, 60–68.

212. Liu, Z., Sun, X.M., Nakayama-Ratchford, N., and Dai, H.J. (2007) Supramolecular chemistry on water-soluble carbon nanotubes for drug loading and delivery. *ACS Nano*, **1**, 50–56.

213. Pantarotto, D., Partidos, C.D., Hoebeke, J., Brown, F., Kramer, E., Briand, J.P., Muller, S., Prato, M., and Bianco, A. (2003) Immunization with peptide-functionalized carbon nanotubes enhances

virus-specific neutralizing antibody responses. *Chem. Biol.*, **10**, 961–966.

214. Wu, H.C., Chang, X.L., Liu, L., Zhao, F., and Zhao, Y.L. (2010) Chemistry of carbon nanotubes in biomedical applications. *J. Mater. Chem.*, **20**, 1036–1052.

215. Zhang, Z., Yang, X.Y., Zhang, Y., Zeng, B., Wang, Z.J., Zhu, T.H., Roden, R.B.S., Chen, Y.S., and Yang, R.C. (2006) Delivery of telomerase reverse transcriptase small interfering RNA in complex with positively charged single-walled carbon nanotubes suppresses tumor growth. *Clin. Cancer Res.*, **12**, 4933–4939.

216. Liu, Z., Winters, M., Holodniy, M., and Dai, H.J. (2007) siRNA delivery into human T cells and primary cells with carbon-nanotube transporters. *Angew. Chem. Int. Ed. Engl.*, **46**, 2023–2027.

217. Guiseppi-Elie, A., Lei, C.H., and Baughman, R.H. (2002) Direct electron transfer of glucose oxidase on carbon nanotubes. *Nanotechnology*, **13**, 559–564.

218. Besteman, K., Lee, J.O., Wiertz, F.G.M., Heering, H.A., and Dekker, C. (2003) Enzyme-coated carbon nanotubes as single-molecule biosensors. *Nano Lett.*, **3**, 727–730.

219. Lin, Y.Q., Zhu, N.N., Yu, P., Su, L., and Mao, L.Q. (2009) Physiologically relevant online electrochemical method for continuous and simultaneous monitoring of striatum glucose and lactate following global cerebral ischemia/reperfusion. *Anal. Chem.*, **81**, 2067–2074.

220. Wang, J. (2005) Carbon-nanotube based electrochemical biosensors: a review. *Electroanalysis*, **17**, 7–14.

221. Liu, J., Chou, A., Rahmat, W., Paddon-Row, M.N., and Gooding, J.J. (2005) Achieving Direct Electrical Connection to Glucose Oxidase Using Aligned Single Walled Carbon Nanotube Arrays *Electroanalysis*, **17**, 38–46.

222. Yu, X., Munge, B., Patel, V., Jensen, G., Bhirde, A., Gong, J.D., Kim, S.N., Gillespie, J., Gutkind, J.S., Papadimitrakopoulos, F. *et al.* (2006) Carbon nanotube amplification strategies for highly sensitive immunodetection of cancer biomarkers. *J. Am. Chem. Soc.*, **128**, 11199–11205.

223. Wohlstadter, J.N., Wilbur, J.L., Sigal, G.B., Biebuyck, H.A., Billadeau, M.A., Dong, L.W., Fischer, A.B., Gudibande, S.R., Jamieson, S.H., Kenten, J.H. *et al.* (2003) Carbon nanotube-based biosensor. *Adv. Mater.*, **15**, 1184–1190.

224. Mattson, M.P., Haddon, R.C., and Rao, A.M. (2000) Molecular functionalization of carbon nanotubes and use as substrates for neuronal growth. *J. Mol. Neurosci.*, **14**, 175–182.

225. Nguyen-Vu, T.D.B., Chen, H., Cassell, A.M., Andrews, R.J., Meyyappan, M., and Li, J. (2007) Vertically aligned carbon nanofiber architecture as a multifunctional 3-D neural electrical interface. *IEEE Trans. Biomed. Eng.*, **54**, 1121–1128.

226. Matsumoto, K., Sato, C., Naka, Y., Kitazawa, A., Whitby, R.L.D., and Shimizu, N. (2007) Neurite outgrowths of neurons with neurotrophin-coated carbon nanotubes. *J. Biosci. Bioeng.*, **103**, 216–220.

227. Thompson, B.C., Chen, J., Moulton, S.E., and Wallace, G.G. (2010) Nanostructured aligned CNT platforms enhance the controlled release of a neurotrophic protein from polypyrrole. *Nanoscale*, **2**, 499–501.

228. Hayashi, T., Kim, Y.A., Natsuki, T., and Endo, M. (2007) Mechanical properties of carbon nanomaterials. *Chem. Phys. Chem.*, **8**, 999–1004.

229. Aoki, K., Usui, Y., Narita, N., Ogiwara, N., Iashigaki, N., Nakamura, K., Kato, H., Sano, K., Kametani, K., Kim, C. *et al.* (2009)

A thin carbon-fiber web as a scaffold for bone-tissue regeneration. *Small*, **5**, 1540–1546.

230. Saffar, K.P., Arshi, A.R., JamilPour, N., Najafi, A.R., Rouhi, G., and Sudak, L. (2010) A cross-linking model for estimating Young's modulus of artificial bone tissue grown on carbon nanotube scaffold. *J. Biomed. Mater. Res. Part A*, **94A**, 594–602.

231. Shao, S.J., Zhou, S.B., Li, L., Li, J.R., Luo, C., Wang, J.X., Li, X.H., and Weng, J. (2011) Osteoblast function on electrically conductive electrospun PLA/MWCNTs nanofibers. *Biomaterials*, **32**, 2821–2833.

232. Overney, G., Zhong, W., and Tomanek, D. (1993) Structural Rigidity and low-frequency vibrational-modes of long carbon tubules. *J. Phys. D-Atoms Mol. Clusters*, **27**, 93–96.

233. Ruoff, R.S. and Lorents, D.C. (1995) Mechanical and thermal properties of carbon nanotubes. *Carbon*, **33**, 925–930.

234. Yakobson, B.I., Brabec, C.J., and Bernholc, J. (1996) Nanomechanics of carbon tubes: instabilities beyond linear response. *Phys. Rev. Lett.*, **76**, 2511–2514.

235. Nardelli, M.B., Yakobson, B.I., and Bernholc, J. (1998) Mechanism of strain release in carbon nanotubes. *Phys. Rev. B*, **57**, R4277–R4280.

236. Lu, J.P. (1997) Elastic properties of carbon nanotubes and nanoropes. *Phys. Rev. Lett.*, **79**, 1297–1300.

237. Treacy, M.M.J., Ebbesen, T.W., and Gibson, J.M. (1996) Exceptionally high Young's modulus observed for individual carbon nanotubes. *Nature*, **381**, 678–680.

238. Wong, E.W., Sheehan, P.E., and Lieber, C.M. (1997) Nanobeam mechanics: elasticity, strength, and toughness of nanorods and nanotubes. *Science*, **277**, 1971–1975.

239. Chopra, N.G. and Zettl, A. (1998) Measurement of the elastic modulus of a multi – wall boron nitride nanotube. *Solid State Commun.*, **105**, 297–300.

240. Falvo, M.R., Clary, G.J., Taylor, R.M., Chi, V., Brooks, F.P., Washburn, S., and Superfine, R. (1997) Bending and buckling of carbon nanotubes under large strain. *Nature*, **389**, 582–584.

241. Iijima, S., Brabec, C., Maiti, A., and Bernholc, J. (1996) Structural flexibility of carbon nanotubes. *J. Chem. Phys.*, **104**, 2089–2092.

3
Organic Conducting Polymers

The discovery that organic polymers that are conjugated (repeating single and double bond units within the polymer structures) can be rendered electronically conducting by doping with appropriate ions revolutionized materials science and resulted in the Nobel Prize being awarded to Alan MacDiarmid, Alan Heeger, and Hideki Shirakawa in the year 2000 [1]. Doping involves the addition of electrons (reduction reaction) or the removal of electrons (oxidation reaction) from the polymer, with concomitant inclusion of counterions to balance the charge subsequently generated on the polymer backbone. The doping process increases the conductivity of the polymer many orders of magnitude.

The most practically useful (environmentally stable) conducting polymers that have emerged from the past three decades of study are based on polypyrrole (PPy) (**1**), polythiophenes (PTh) (**2**), and polyaniline (PANi) (**3**). It turns out that these dynamic materials have properties that provide a unique perspective on how the bionic interface might be controlled and developed.

(1) Polypyrrole

(2) Polythiophene

Organic Bionics, First Edition. Gordon G. Wallace, Simon E. Moulton, Robert M.I. Kapsa, and Michael J. Higgins.
© 2012 Wiley-VCH Verlag GmbH & Co. KGaA. Published 2012 by Wiley-VCH Verlag GmbH & Co. KGaA.

The number of monomer units per unit positive charge is usually two to four for PPy and PTh. A$^-$ is the dopant anion.

(3) Polyaniline

In PANi, $y = 1$ for leucoemeraldine, $y = 0.5$ for emeraldine, and $y = 0$ for pernigraniline; m determines the molecular weight.

Some derivatives of PTh, in particular poly(3,4-ethylenedioxythiophene) (PEDOT) (4), and PANi, in particular, the methoxy sulfonated aniline poly(2-methoxyaniline-5-sulfonic acid) (PMAS) (5), have attracted attention mainly due to the processability of these materials in aqueous media.

(4) Poly(3,4-ethylenedioxythiophene) (PEDOT)

(5) Polymethoxyaniline sulfonate (PMAS)

The diverse array of tunable properties available with organic conducting polymers (OCPs) is based primarily on their composition and doping level [2]. In the case of PPy and PTh, the reduced (undoped) neutral form is insulating, while the oxidized (doped) form is conducting. In the case of PANi, the fully reduced form ($y = 1$) or fully oxidized form ($y = 0$) are insulating, while the partially oxidized form ($y = 0.5$) is the most conducting. The conjugated polymer backbone obviously affords electronic and electrochemical properties of interest. As in the case of PEDOT and PMAS, the chemical and physical properties of these polymers can be modified by covalent attachment of appropriate functional groups to the polymer backbone. The molecular dopant anion (A$^-$) can range from a very simple species, such as a chloride anion, to a complex biomolecular species, such as a protein or a biological polyelectrolyte. A further degree of control over properties is provided by manipulation of the structure at the nano- to microdomain. Such manipulation provides routes to improve the conductivity and mechanical strength of the material. Interestingly, how a surface might lend itself to chemical or biological modification is also dependent on nanostructure. A final dimension of control over chemical and physical properties lies in the fact that they can be tuned *in situ* via the redox state of the materials. This property affords a dimension not available in other types of materials and could prove critical in constructing the ideal interface for a medical bionic device. The ability to control properties such as the modulus, ion permeability, and surface energy *in situ* has given rise to a number of potential applications of relevance to the complex area of medical bionics.

OCPs are usually formed by the oxidation of monomer by either electropolymerization at a conductive substrate (electrode) through the application of an external potential or a current or chemical polymerization by the use of an appropriate chemical oxidant. These approaches produce OCP materials with different forms: chemical oxidations generally produce powders, while electrochemical synthesis leads to films deposited on the working electrode. Synthesis by exposing monomer vapor to an appropriately cast oxidant can result in adherent film formation with a protocol denoted "vapor phase polymerization" (VPP).

3.1
Polypyrrole

For medical bionics applications, one would prefer to assemble the OCP in aqueous media and at moderate potentials in order to maintain the viability of any bioactive entity to be incorporated into the polymer structure. The formation of PPy is possible under such conditions.

PPy can be formed by the oxidation of pyrrole at a suitable anode in an electrochemical cell to form an insoluble, conducting polymeric material as a deposit on the anode. The polymerization reaction can be represented simply as

$$(3.1)$$

(n = degree of doping 3–4 and m relates to the polymer chain length, which determines molecular weight).

In Eq. (3.1), A^- is a counterion necessary to balance the charge on the doped polymer backbone. The counterion content is high and can even be greater than 50% w/w. The incorporation of an appropriate dopant forms a major constituent of the composition and hence the inherent chemical–biological properties.

For medical bionics applications, the media used to support electropolymerization is chosen so as to deliver an OCP with appropriate composition (usually containing biologically relevant molecules), electronic properties, mechanical properties, and surface morphology. The use of pyrrole, as opposed to thiophene and aniline, provides some advantages since this monomer is water soluble and thus presents minimal hazards to denature biological molecules during synthesis. In addition, aqueous solvents are preferable to organic solvents from the point of view of cost, ease of handling, safety, and the range of counterions (dopants), including biomolecular dopants, that can be used.

The electrolyte salt to be used will have a significant effect on the final OCP properties since this is the source of the molecular dopant. For example, when polyelectrolytes are incorporated as dopants [3], conducting polymers with very high water content (electronic hydrogels) are usually obtained. Formation of an open hydrophilic polymer network encourages continued growth of each domain since the polymer prefers to grow and deposit in the more hydrophobic regions of the structure, ensuring continued phase segregation during growth. The following chemical structures are examples of biological polyelectrolytes that have been incorporated into conducting polymers as the molecular dopant. Chondroitin sulfate (CS) (6), dextran sulfate (DS) (7) [4], hyaluronic acid (HA) (8) [5], or heparin (9) [6] is readily incorporated into PPy during electrosynthesis.

(6) CS – Chondroitin sulfate
$R_1, R_2, R_3 \equiv -SO_3Na$

(7) DS – Dextran sulfate
$R \equiv -SO_3Na$

(8) HA – Hyaluronic acid

(9) Heparin

n = number of repeat units and defines the molecular weight

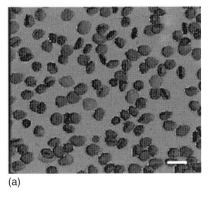

(a)

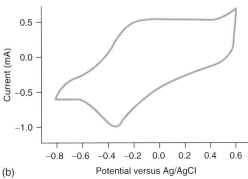

(b)

Figure 3.1 (a) A polymer film fixed in glutaraldehyde (0.15%)/phosphate buffer immediately after synthesis. Scale bar: 10 µm. (b) Cyclic voltammogram of polymers containing erythrocytes cycled with and without the presence of anti-Rh(D). Polymers were cycled between +0.5 V and −0.8 V (vs Ag/AgCl) beginning at +0.17 V, in a solution of NaCl (0.15 M), at a scan rate of 40 mV s^{-1}. (Adapted with permission from Campbell *et al.* [9].)

Other biopolymers such as oligonucleotides [7] and even DNA [8] have been incorporated into PPy as dopants, presenting the exciting possibility of their future use in gene-modulating applications.

An open porous hydrogel network is created by incorporating synthetic polyelectrolytes such as polystyrene sulfonate (PSS) as the dopant. The excess charge introduced by incorporating high charge density PSS chains results in a highly hydrophilic swollen gel. Polymerization of pyrrole from an electrolyte containing the polyelectrolyte PSS and red blood cells results in an OCP with intact viable cells incorporated throughout the resulting organic [9] conducting matrix (Figure 3.1).

An interesting option in terms of dopants is to incorporate an organic conducting polyelectrolyte such as sulfonated PANi (PMAS) as the molecular dopant (Scheme 3.1). Formation of an electrically conducting hydrogel (containing >90% (w/w) H$_2$O) with multiple electrochemical switches is the result [10].

This polymer is noncytotoxic to L929, primary neuronal cells, and transformed neuronal cells and has been shown to promote neuronal growth and neuroglial cell migration on electrical stimulation [11].

Electropolymerization is most suitable for coating other conducting materials such as metals or carbon materials since the polymer produced is usually insoluble and deposits on these conductors in a manner akin to electroplating. The simplest means of inducing the polymerization process for PPy is to apply a sufficiently positive constant potential. The potential chosen will influence the rate of oxidation and, therefore, polymerization. If the rate of polymerization is too slow (typically at potentials less than +0.50 V vs Ag/AgCl reference), oxidation of the pyrrole monomer may occur without

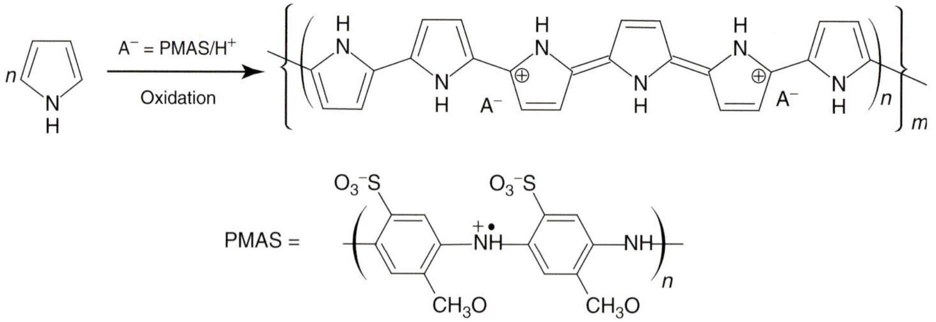

Scheme 3.1 Polymerization of pyrrole in the presence of the sulfonated conducting polymer PMAS provides a multifunctional electronic hydrogel.

deposition since the solubility limit will not be exceeded before the products diffuse away from the reaction zone near the electrode surface. However, the upper value of the potential is limited by a process that results in overoxidation of the polymer. Polymer overoxidation results in a less conductive and more porous polymer product [12, 13] with inferior mechanical properties. However, when incorporating biological entities during polymerization, the upper potential limit will more likely be determined by the range in which the bioentity is electrochemically stable. The accurate control of potential is critical in maintaining the viability of biological entities to be integrated.

An alternative is to apply a constant current to drive the reaction. This provides more accurate control over the rate of polymerization. The current density required for successful polymerization varies greatly depending on the dopant species to be incorporated with the monomer. For example, current densities as low as $0.25 \, \text{mA cm}^{-2}$ have been used to polymerize PPy with the biological dopant HA [14], while polymerization of PPy with the dopant *para*-toluenesulfonic acid (*p*TS) [15] has been performed at $2.0 \, \text{mA cm}^{-2}$. Again, if the rate is too slow, oxidation without deposition may occur. However, if the rate is too high, the potential may stray into the region where overoxidation of the polymer or oxidative damage to biological entities occurs.

The nature of the conducting material determines how easily the monomer can be oxidized. It also determines the degree of accumulation of the reactants, intermediates, and the final polymeric product. For example, some oxide-containing electrodes such as aluminum undergo covalent binding reactions with the polymer to form an extremely adherent film. In metals that are themselves easily oxidized (e.g., Al and Cu), addition of electrocatalysts such as Tiron results in a significant decrease in the polymerization potential.

For bionic applications, it is important to be able to prepare conducting polymers in a range of different forms. For example, the use of longer lengths of conducting polymer (CP) fibers of either microdimensions or nanodimensions is appropriate for providing electrical stimulation and topographical cues to control the direction of cell growth. Conducting polymer gels provide a 3D network for growth of living cells.

These structures can be realized by polymerization on or through a host structure such as a preformed fiber or gel. Alternatively, they can be formed by wet spinning (to produce microdimensional fibers), electrospinning (to produce nanodimensional fibers), extrusion or inkjet printing (to produce microdimensional features), or even dip-pen nanolithography (to produce nanodimensional features).

These fabrication techniques usually involve the use of OCP dispersions. It is possible to use electropolymerization to form soluble conducting polymers such as PMAS, micrometer-sized colloidal dispersions [16], and even stable nanodispersions [17]. The latter is achieved by coating polyurethane nanoparticles.

Of particular significance to the bionics community is the fact that proteins [18], polyelectrolytes [19], and even dopants that produce chiral nanoparticles have been successfully incorporated into these colloidal dispersions [20].

Chemical polymerization, wherein the monomer is exposed to a chemical oxidant, has also been widely used to produce micrometer-sized colloidal dispersions or nanodispersions. A range of water-soluble steric stabilizers such as poly(vinylalcohol), poly(ethyleneoxide), or poly(vinylpyridine) have been used to produce stable colloidal dispersions [21, 22]. Methyl cellulose has also been used [23] as have cationic and anionic polyelectrolytes. The most widely used chemical oxidants have been ammonium persulfate, $FeCl_3$, and iron *para*-toluenesulfonate.

One approach to the formation of PPy and other OCP nanoparticles is through the use of micellar polymerization and microemulsion techniques. The advantage of such an approach is that the particle size can be predefined by establishing the appropriate size and geometry of the templating micelle. Using this approach, Jang and coworkers [24] have reported the synthesis of PPy nanoparticles of 2.0 nm in size.

Chitosan has been used as a molecular template to promote formation of nanosized PPy particles on silica [25]. Interestingly, another biomolecule (heparin) has been shown to have a similar templating effect. In the case of heparin, it acts as both the molecular dopant and a (nano)structure-directing template [26]. Others [27] have used lipid tubules as templates during chemical oxidation of pyrrole to form nanofibers with diameters between 10 and 50 nm and lengths reaching up to several hundred micrometers.

Using PMAS as the dopant for PPy, stable nanodispersions have been formed without the need for additional stabilizers [16]. The PMAS dopant introduces multiple electrochemical switching pathways to the resultant material, which may have benefit for the areas of application of interest here.

The above dispersions can be used to coat biomedical materials that are nonconductors or metals that are highly active and, therefore, corrode in even mild oxidative conditions. An alternative approach to coating new conductive substrates is the use of VPP.

Vapor phase *in situ* polymerization (deposition) has been employed, for example, by passing pyrrole vapor over cotton thread coated with the oxidant $FeCl_3$ [28]. Typically, the surface resistance of such conducting-polymer-coated fabrics decreases with increasing polymer deposition. It has been shown that conducting PPy can be formed by exposing a layer of the oxidant salt (e.g., $FeCl_3$) to pyrrole monomer in the vapor phase [29]. Adherent conducting polymer films can be produced in this way. Interestingly, this VPP method gives rise to a highly swollen polymer structure containing excess oxidant that is readily removed using a washing step [30]. Including functional molecules, such as organometallic catalysts, enzymes [31], or even neurotrophins [32], in the wash solution provides a convenient way to load the conducting polymers with functional molecules.

3.2
Polythiophenes

The preparation of PThs proceeds as for PPys via oxidative polymerization in the appropriate manner. However, the formation of conducting PTh has been thwarted by the fact that at potentials required to oxidize the thiophene monomer, the polymer itself becomes overoxidized [33] to a less conducting form. The use of alkyl thiophenes such as 3-methylthiophene [34] also overcomes this problem, as the oxidation potential of the monomer can be significantly lowered by the addition of electron-donating substituents. PThs can also be derived from oxidation of the bithiophene or terthiophene monomers alone [35, 36]. These are again oxidized at a lower potential than the thiophene monomer. We have shown that it is possible to control the oxidation potential of thiophenes over a wide potential range by addition of either electron-withdrawing or electron-donating groups.

Alternatively, small amounts of bi- or terthiophene are added to reduce the polymerization potential [37].

A range of alkoxy groups [38] (structures **10–12**) has also been added to the bithiophene starting material in order to reduce the oxidation potential.

(10)   **(11)**   **(12)**

PEDOT (**10**) has attracted the most attention for medical bionics applications because of the high conductivities attainable and the possible advantages of improved stability. It is worth noting, however, that the issue of stability in a medical bionic device context really pertains to the ability of the conductor implanted to induce formation of a stable biology–electronic interface where that is required. As with PPy, a number of bioactive dopants including collagen [39], heparin [40], and choline oxidase [41] have been incorporated into PEDOT during synthesis. The ethylene dioxythiophene (EDOT) monomer is not highly water soluble; however, the use of a hydroxyl methylated EDOT enables monomer concentrations of 0.1 M to be used during electrochemical synthesis, facilitating the incorporation of peptides [42].

The use of organic solvents is usually preferred for thiophene oxidation since the monomer is more soluble in this media. The presence of water during polymerization causes deterioration in polymer properties [43]. This limits the use of biofunctional molecules to those that can withstand suspension in organic solvents and retain their functionality.

However, other workers have shown that polymerization of functional thiophenes from aqueous media is possible if surfactants [44, 45] are used to help solubilize the monomer. In the case of sodium dodecyl sulfate (SDS) being used as a solubilizing agent for bithiophene [46], it was also found to lower the oxidation potential. For EDOT [47] and other copolymers containing alkoxythiophene and bithiophene groups [44], a similar effect was observed. This water solubility and lowering of the monomer oxidation potential facilitate the development of protocols that allow incorporation of biomolecules during polymerization.

The chemistries available on PThs are much more versatile than on PPys. Consequently, the options available in producing processable (soluble or stable dispersions) formulations are much greater. As with PPy, PTh can be produced using a chemical oxidant. However, because of the limited solubility of thiophene, this reaction is usually carried out in nonaqueous media. Copper(II) perchlorate has been used as an oxidizing agent in acetonitrile to provide doped PTh material [48] with a conductivity of 8 S cm^{-1}. Others [49] have used a chemical polymerization process that ensures the production of well-defined head-to-tail coupled poly(3-alkylthiophenes) with increased conductivity. The polymerization method of VPP is also a popular synthesis method for the functionalized thiophene PEDOT that produced adherent films [50, 51]. PEDOT can also be polymerized from an aqueous environment using the micellar route described previously. PEDOT nanoparticles with diameters in the range 35–100 mm have been prepared with conductivity of ~50 S cm^{-1} have been prepared [52].

The dopant A$^-$ incorporated into the polymer is usually the conjugate base from the acid. As with other OCPs, the nature of the dopant anion is critical in determining the morphology [53], conductivity [54], and subsequent switching characteristics of the polymer.

3.3
Polyanilines

As with PPys and PThs, PANi can be directly electrodeposited onto a conducting surface. This does require the use of an acidic electrolyte to solubilize the aniline monomer and maintain the polymer in the conducting form during the deposition process. With respect to coating nonconducting surfaces, the ability to create soluble or dispersible formulations is required.

Stable colloidal dispersions of PANi have been produced electrochemically [52]. However, a much more common approach is via chemical oxidation. The most widely employed chemical oxidant has been aqueous ammonium persulfate, $(NH_4)_2S_2O_8$, leading to the incorporation of HSO_4^-/SO_4^{2-} as the dopant anions (A$^-$) in the PANi/HA product. The use of FeCl$_3$ also enables polymerization to be carried out in polar organic solvents such as methanol rather than water [55].

The major route to PANi colloidal dispersions has been through the chemical oxidation of the monomer in the presence of polymeric steric stabilizers and electrosteric stabilizers (polyelectrolytes), such as poly(vinylalcohol), poly(ethylene oxide), PSS, and DS [56]. It has been found that the stabilizer can act simultaneously as a dopant, imparting new functionality to the polymer or additional compatibility for the final application. Formation of PANi nanoparticles has been achieved via polymerization in micelles using either SDS or dodecylbenzenesulfonic acid (DBSA) [57] as the surfactant stabilizer. Particle sizes in the range of 10–30 nm with conductivities as high as 24 S cm^{-1} have been reported.

Interfacial polymerization provides a facile and versatile method to produce emeraldine salts (ESs) as nanofibers with diameters between 30 and 120 nm, depending on the nature of the dopant anion employed [58, 59]. Li and Kaner [60] demonstrated that homogeneous nucleation of PANi results in nanofibers.

3.4
Properties of OCPs

3.4.1
Conducting and Electrochemical Switching Properties

Conductivity, the ability to transport charge, electrochemical switching properties, and the ability to transfer charge are properties of paramount importance for bionic applications. PPys have conductivities typically in the range $1-400\,S\,cm^{-1}$. The conductivity attainable is highly dependent on the dopant anion utilized. With biologically relevant dopants such as proteins or biopolymers, the conductivity tends to be at the lower end of the above range. These values, however, are still sufficient to allow direct electrochemical stimulation of living cells and/or effective electrochemical switching of the PPy backbone, at least at moderate frequencies $(1-10\,Hz)$. This frequency range can be increased by use of smaller electrodes.

The presence of alkyl functional groups on thiophene monomers [61] can be used to advantage in preventing overoxidation of the PThs during synthesis, thus enhancing conductivity. Conductivities as high as $7500\,S\,cm^{-1}$ have been obtained for polymethylthiophene [34]. As with PPy, the counterion used during electropolymerization influences the conductivity of PTh [62, 63]. We have recently shown that spontaneous dedoping of a number of functionalized PTh readily occurs in phosphate buffer.

PANi has an electronic conduction mechanism that is unique among the known conducting polymers, as it is doped by protonation as well as the p-type doping described for PPy. This results in the formation of a nitrogen base salt rather than the carbonium ion of other p-doped polymers [64]. Therefore, the conductivity of PANi depends on both the oxidation state of the polymer and its degree of protonation. The maximum conductivity occurs when PANi is 50% doped by protons. At doping levels higher than 50%, some amine sites are protonated, and at levels lower than this, some imine sites remain unprotonated. This characteristic greatly diminishes the practical applicability of PANis in bionic applications.

3.4.2
Electrochemical Switching Properties

Several parameters influence the electrochemical switching process, the most relevant to medical bionics applications being

- the composition of the polymer, and
- the operational environment.

Charge can be reversibly added to or removed from an OCP by cycling the material through its oxidized and reduced states. As the switch from oxidized to reduced state occurs, there is a concomitant decrease in the conductivity. The process for PPy may be described by Eq. (3.2).

Contracted state

$$+ 2e^- - 2\,X^+ \qquad\qquad - 2e^- + 2\,X^+ \qquad\qquad\qquad (13)$$

Expanded state

$$(3.2)$$

where A^- = anions, e^- = electrons, and X^+ = cations.

A typical cyclic voltammogram arising from this process, in which the potential (voltage) is swept while recording the current, is shown in Figure 3.2.

Medical bionics applications often require the incorporation of molecules into the OCP that result in biocompatible and/or bioactive structures. It is important that such incorporation does not overly compromise conductivity and/or the electrochemical switching process. This conductivity and electrochemical switching behavior must also be retained in biologically relevant electrolytes.

The operational environment for the development of medical bionic OCPs initially involves primary evaluation in appropriate biological buffers or cell culture media for *in vitro* experiments, leading ultimately to the biological (tissue) environment of interest for *in vivo* work.

The incorporation of electroactive dopants introduces multiple switching capabilities since the dopant itself can be oxidized/reduced. For example, a sulfonated ferrocene has been introduced into PPy [65]. This is particularly interesting for medical bionics applications, as it provides an additional electron transfer mediation pathway to biomolecules such as redox active proteins.

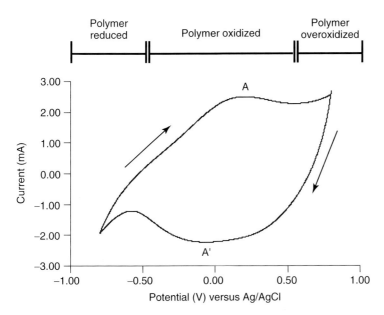

Figure 3.2 Cyclic voltammogram of PPy polymer containing the neurotrophin NT-3 as the dopant; recorded in 0.2 M phosphate-buffered saline (PBS) (pH 7.4) between −0.8 V and +0.8 V at a scan rate of 100 mV s^{-1}. Arrows indicate the direction of potential scan. (Adapted with permission from Thompson and coworkers [15].)

During the reduction of PPy that contains large counterions, such as DS or CS, dopants that are considered immobile and are unable to be removed and cations are consequently incorporated as part of the reduction process (Eq. (3.2)).

The cation is derived from the operational environment in which the polymer is reduced, such as K$^+$ or Na$^+$ from a phosphate buffer solution. Such movements of cationic species have an inherent potential to affect biological processes in themselves and need to be accounted for in the design of implantable bionic devices.

The incorporation of a redox active conducting polymer as a molecular dopant provides yet another dimension to the electrochemical switching capabilities since multiple electronically conducting pathways are provided (Figure 3.3) [11].

The size of the electrode also affects switching properties. The ability to switch at high speeds has been demonstrated previously by other workers using microelectrodes [66]. Conversely, larger electrodes will be slower to switch and the degree of switching (percentage of material affected) will be much less efficient.

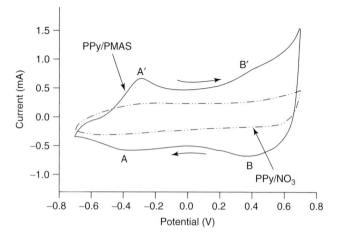

Figure 3.3 Cyclic voltammograms of PPy/PMAS and PPy/NO$_3$ films obtained in 0.1 M NaNO$_3$ at a scan rate of 100 mV s^{-1}. The arrows indicate the direction of potential scan. The cyclic voltammetry (CV) of PPy/PMAS exhibits a well-defined redox couple at −0.40 and −0.30 V (A/A′), attributed to the PPy oxidation/reduction, and additional redox responses at +0.35 and +0.40 V (B/B′) (vs Ag/AgCl reference electrode) attributed to the oxidation/reduction of PMAS. (Adapted with permission from Liu and coworkers [11].)

3.4.2.1 Polythiophenes

As with PPy, the properties of PTh can be modulated by application of electrical stimuli, according to Eq. (3.3).

Contracted state

$$+2e^- - 2A^- \qquad\qquad -2e^- + 2A^- \qquad\qquad\qquad \textbf{(14)}$$

Expanded state

$$A^- \qquad\qquad\qquad\qquad A^-$$

(3.3)

where A^- = anions and e^- = electrons.

This simplistic description belies the reality. Different sites are switched at different potentials, and when A^- is relatively immobile, cation movement accommodates the oxidation/reduction process.

The switching properties are determined by the counterion incorporated and the substituents attached to the thiophene ring (Table 3.1) [67]. The electrolyte used for switching also has a marked effect on the potential, rate, and reversibility of the switching process [68].

PThs are, in general, more hydrophobic materials than PPy. Electrochemical switching is, therefore, not so efficient in aqueous media as encountered in the biological applications of interest here. As a result, some workers have used the ability to attach functional groups such as polyethers [67] to PTh in order to increase their hydrophilicity and improve electrochemical behavior in aqueous solutions.

The switching characteristics are also different, in that PThs are generally more difficult to oxidize and overoxidize. However, the fact that PThs must normally be prepared from nonaqueous media limits the range of dopants (particularly functional biological molecules) that can be incorporated.

Despite the lack of electrochemical switching in aqueous media [69], PThs are still capable of electron transfer as evidenced by the ability to oxidize–reduce $Fe(CN)_6^{4-/3-}$ [69] in aqueous media. This shows that PThs can provide an OCP platform capable of injecting/extracting electrons without the normal changes in chemical and physical properties associated with the oxidation/reduction of the polymer backbone.

PANis undergo two distinct electrochemical switching processes transiting the fully reduced, partially oxidized, and fully oxidized states. It is the partially oxidized state that is the most highly conducting.

The rate of switching between these states depends on the electrolyte used [70] (switching is faster with smaller ions) and the solvent employed [71]. If the PANi polymer is exposed to potentials above the second oxidation process, irreversible degradation occurs [72].

When polyelectrolytes are incorporated as dopants in PANi, this enables the electrochemistry to be carried out on PANi even in neutral solutions [73], which has significant implications on their use in biological applications. While, in general, physiological pH ranges between 7.3 and 7.4, there are some tissues such as the stomach, skin, kidney, and pancreas where pH differs markedly from this range. Any pH-dependent switching from conducting to nonconducting states can be employed to control the release of biofactor/dopant from the OCP in response to specific changes in pH. Otherwise, the choice of conducting polymer must take into account the pH of the intended *in vivo* environment, in order that the desired functionality, particularly electroactivity, is maintained.

Table 3.1 Polythiophenes.

Polymer (ClO$_4^-$)	A	C	F	D	E	B
E_{onset} (V) (vs Ag/Ag$^+$)	−0.54	−0.22	−0.02	0.08	0.40	0.43
Water contact angle	83.6 ± 4.8°	104.9 ± 2.7°	34.9 ± 1.6°	90.0 ± 2.5°	84.8 ± 6.4°	41.3 ± 2.5°

3.5
Chemical-Biological Properties

Arising from polymer-dopant combination effects, polymer surface chemistry has been shown to exert a significant influence on the interaction of these materials with biomolecular entities such as cell adhesion molecules. For example, variation of the surface-presenting functional group has been shown to affect the tertiary conformation of fibronectin adsorbed to self-assembling alkanethiol monolayers [74]. The initiation and patterning of growth behavior in cells (e.g., neurons) that compose biological tissues arises from several environmental influences, of which a critical factor is the permissive interaction with the substratum [75]. As such, the extensively dynamic chemical properties of OCPs are capable of modulating the biological behavior of cells.

3.5.1
Polypyrroles

Even simple PPys (with incorporated counterions such as chloride or nitrate) are inherently versatile molecular structures capable of undergoing a range of molecular interactions. They are ion exchanger materials [76, 77], which are also capable of undergoing hydrophobic interactions. The counterion employed during synthesis can have a marked effect on the anion exchange behaviors [78]. Polyelectrolytes may also be incorporated as counterions during synthesis [79–81], and since they would be difficult to remove, the anion exchange capacity is reduced as the cation exchange capacity concomitantly increases.

It is also possible to incorporate a wide range of chemically functional counterions into PPy [82, 83], largely afforded by the fact that pyrrole can be polymerized from aqueous solutions. In cases where a counterion with a specific activity is incorporated, the polymer may act as no more than a carrier or a means of electronic communication. Numerous examples have been reported, where functional biomolecules such as antibodies [84–87], enzymes [88–92], anti-inflammatory agents (e.g., dexamethasone) [93, 94], and neurotrophins [95, 96] have been incorporated into PPy. These exciting studies highlight the potential for PPy to be used as an effective polymer for medical bionics applications.

Polynucleotides [7, 97] and DNA [8] have also been incorporated in an attempt to harness the specific properties of these molecules. To date, however, these species have been useful for the specification of physicochemical properties of the polymer, rather than as bioactive species that are released to promote biospecific effects on tissue systems. Ongoing studies will ultimately provide the rationale for the incorporation of bioactive nucleic acids, such as antisense or small inhibitory nucleic acids, that promote the regulation of biological activity at the gene and transcript levels. We have successfully incorporated mammalian red blood cells during electrosynthesis, using polyelectrolytes as the codopants [9].

The activity of incorporated functional molecules such as proteins can be altered by the application of appropriate electrical stimuli [84, 85, 98, 99]. Initial forays into this area involved [84] the incorporation of proteins into phases suitable for affinity chromatography. Later studies involved flow injection analysis using an antibody-loaded conducting polymer as the sensing element [85]. Antibodies to a number of species have now been incorporated and the ability to control the Ab–Ag interaction demonstrated [86, 87, 98–100].

3.5.2
Polythiophenes

As with PPy, a hydrophobic backbone is formed and the polymer has ion exchange properties. Modification of chemical properties by incorporation of appropriate counterions is not so accessible since polymerization must be carried out from nonaqueous solution and occurs at more anodic potentials compared to pyrrole.

However, the fact that thiophenes are easier to derivatize than pyrroles is attractive, in that chemical functionality can easily be introduced. A range of functional groups have been introduced [101–106] to modify the chemical properties of PTh. The ability to attach active groups such as amino acids provides an attractive route for modifying the molecular recognition properties of the resultant polymer. The decreased activity of the sulfur groups compared with the NH group also means that derivatization after polymerization is more easily accomplished [107–109].

3.5.3
Polyanilines

The strongest chemical interactions of PANi arise through its anion exchange properties, and these properties differ in several ways from those of conventional ion exchange resins. This may be attributed to charge delocalization, as suggested by charge-configuration-related phenomena observed in studies of amino acid interactions with PANi [110]. These preliminary studies suggest that specific protein–polymer interactions may be regulated according to the protein motif(s) presented to the interactive molecular componentry of the polymer scaffold.

Incorporation of bioactive molecules into PANi is not so readily achieved since electropolymerization must normally be carried out at low pH, and this normally denatures proteins. However, thin polymeric coatings containing enzymes have been produced by polymerization from solutions buffered to physiological pH (pH 7) [111, 112]. In other work similar to that undertaken with PPy, enzyme-assisted polymerization was used to produce PANi with DNA as the dopant [113]. Interestingly, the PANi adopts the chiral nature of the DNA strands.

3.6
Mechanical Properties

To remain functional as a component of any medical bionic implant, a material must possess adequate mechanical properties to withstand the stresses and strains that occur during service within the intended environment of application. For example, stand-alone materials must be sufficiently stiff to maintain their shape and sufficiently strong to resist rupture when subjected to tensile forces and pressure gradients. Similarly, coatings may be subjected to high stress resulting from differences in thermal expansion between coating and substrate. Thus, the coating material must be highly ductile, so that it is able to expand and contract without cracking or delaminating. Likewise, materials implanted into living tissue need to be able to maintain their structural integrity within a potentially aggressive aqueous environment with movement and pressure exerted by biological fluids (e.g., blood) and "solid" tissues (e.g., muscle and bone).

The most commonly reported measures of mechanical properties are the Young's modulus (E, a measure of material stiffness), the tensile strength (σ_B), and the percentage of elongation at break (ε_B, a measure of ductility or brittleness).

3.6.1
Polypyrroles

The composition of the polymer (dictated by the polymer backbone and the counterion) and the polymerization conditions have a significant effect on mechanical properties. The tensile strength of PPy films increases as the polymerization temperature decreases [114]. Smoother films formed at lower polymerization temperatures exhibit higher tensile strength and elongation at break [115].

The solvent used during electropolymerization also affects mechanical properties. For example, the incorporation of a small amount of water with acetonitrile as the polymerization solvent improves the mechanical properties of PPy/pTS films [116]. PPy/pTS films formed from propylene carbonate also showed superior mechanical properties in comparison to those formed from acetonitrile solvent [117]. Generally, polymerization from aqueous media, as required for incorporation of intact biocomponents, results in inferior mechanical properties.

The nature of the counterion incorporated into the PPy matrix has a effect on mechanical properties [117–119]. It has been observed, for example, that the pTS^- counterion produced flexible films, whereas other counterions investigated produced brittle films [119].

There is still a need to acquire data on the stability of OCPs in environments relevant to biological applications. Previous studies have shown that blanket statements on "stability" with scant regard to the operational (chemical and electrochemical) environment have limited usefulness.

For example, two aspects of the mechanical properties of PPy films in different environments have been studied in our laboratories [120]: the effects of different salt solutions and the effect of an applied potential. The films become more ductile in chloride solutions and potassium hydroxide solution. Little change in properties is noted in air or water. The sensitivity of the polymer to environmental conditions demonstrates the importance of determining its behavior in the actual service environment, particularly in the case of its intended use for biological applications.

When electrical potentials are applied, significant changes in mechanical properties have been observed [121]. Negative potentials induced a much higher plasticity in PPy/pTS. When tested without an applied potential, the samples showed classic brittle behavior with an elongation at break of 10%. However, when the polymer was reduced, the elongation at break increased to ~20%. The impact of the applied potential on the mechanical properties is further complicated by the electrolyte employed. For example, with divalent cations in the electrolyte, a more negative potential is required to induce ductility. This correlates with the ease of ion transport into the polymer.

3.6.2
Polythiophenes

Mechanical properties of electrochemically prepared PTh have generally been poor [122], although some reports have described the preparation of PTh films that have extremely high tensile strength and modulus [123]; with tensile strength of up to 140 MPa reported. In other work, high tensile strengths (135 MPa), along with very high modulus (46 GPa) and reasonable elongation at break (4%), have been prepared [122].

Others have reported [124] that the reduced, undoped form of PTh had a higher stiffness ($E = 1300$ MPa) and tensile strength ($\sigma = 55$ MPa) than the oxidized, doped PTh ($E = 360$ MPa; $\sigma = 15$ MPa). It has been reported [125] that the mechanical properties (E, σ_B, and ε_B) of electrochemically reduced PTh films were all increased in comparison to the as-prepared, oxidized films.

3.6.3
Polyanilines

The electropolymerized ES form of PANi is highly porous and, consequently, has low mechanical strength. In contrast, the polymer made from solution is much less porous and is widely used as free-standing films and fibers.

It is possible to prepare free-standing films from PANi in the reduced (leucoemeraldine) state, and when oxidized to the emeraldine state the films became brittle. The polymerization potential has also been found to influence the mechanical properties of PANi with a break of 40% achievable under optimal synthesis conditions.

Improvements in the ductility of PANi ES films can be achieved using plasticizing dopants such as di(2-ethylhexyl)ester of 4-sulfophthalic acid (DEHEPSA) [126, 127].

3.7
Surface Morphology

Atomic force microscopy (AFM) has been used to determine surface morphology and roughness of electrochemically prepared PPy films. Li and coworkers [128] showed that a nodular surface arises very early in the electropolymerization process where "microislands" first form on the (gold) electrode surface. With longer polymerization times, the film thickness increases and the surface consists of close-packed nodular grains of submicrometer diameters. Barisci and coworkers [129] found a strong correlation between the surface potential of PPy films and the nodular surface morphology. They concluded that the nodules are dopant-rich, high-conductivity regions where nuclei initially formed and, subsequently, around which polymer preferentially grew.

The influence of dopant type on surface morphology has been studied using AFM by Silk and coworkers [130, 131]. Thin PPy films ($<1 \mu m$ thick) prepared with four different dopants (Cl^-, ClO_4^-, SO_4^{2-}, dodecyl sulfate) were indistinguishable, all consisting of globules $100-300 \, nm$ in diameter and $10-30 \, nm$ in height. With thicker films ($>5 \mu m$ thick), clear differences were observed. While the sulfate- and dodecyl-sulfate-doped films maintained approximately the same surface morphology in thick films as in the thinner films, thick Cl^- and ClO_4^- doped films generated "cauliflower" structures consisting of large protrusions of $\sim 2 \mu m$ in diameter and $0.5-1 \mu m$ in height. These protrusions showed a substructure of the smaller globules seen in thinner films. For all dopant types, the diameter of the smaller globules increased linearly with the square root of film thickness, suggesting a similar growth mechanism for all dopant types. Similar structures and a similar dependence of globule diameter on film thickness

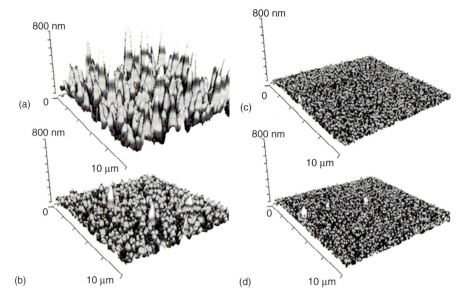

Figure 3.4 AFM images of PPy films doped with biological polyelectrolytes. (a) PPy/pTS, (b) PPy/HA, (c) PPy/CS, and (d) PPy/DS.

have also been reported for PANi films, suggesting that globule formation and growth is common to both electrochemically formed polymers.

More recently, the incorporation of pTS, and other biological polyelectrolytes such as HA, CS, and DS, as dopants in PPy films was found to have a dramatic effect on the surface roughness of their nodular morphologies [4]. AFM imaging revealed that PPy/pTS (100 nm rms) and PPy/HA films (21.8 nm rms) were significantly rougher than PPy/CS (8.2 nm rms) and DS films (7.5 nm rms) (Figure 3.4).

The electrode material employed has also been found to influence the surface morphology of electrochemically prepared PPy films. In our own work, we used transmission electron microscopy (TEM) to investigate the cross-sectional structure of PPy films prepared on different substrates [120]. In all cases, the large surface globules were observed to be the caps of cone-shaped structures that extended to the PPy–electrode interface. Interestingly, Yoon and coworkers [132] showed that the cauliflower-type surface morphology could be virtually eliminated by carefully polishing the electrode surface [132].

In biological applications, topography has been shown to influence the extent and orientation of neuronal growth [133], while other studies have shown that such growth decisions in neurons arise from synergistic effects of topography and cell adhesion molecules attached to the surface [134]. Findings such as these highlight the challenge in identifying the specific (singular) characteristics of OCPs that influence cell behavior. From this perspective, it is even more challenging to identify such factors under electrical stimulation since they are likely to change in character (e.g., swell, actuate, or acquire surface energy) with the passage of electrons along the polymer backbone.

3.8
Conclusions

It is an obvious, albeit "bionically serendipitous," aspect of OCP fabrication that the way in which OCPs such as PPy are assembled has a dramatic effect on their chemical, electrical, mechanical, and, ultimately, their biomimetic (or medical bionic) properties. This information can be used to advantage in material design from the molecular level to specify or "dial up" the characteristics that will ultimately facilitate optimal bionic control of cell systems that come into contact with the polymer matrix. Factors influencing these properties continue to be studied intensively, thereby adding more information to the already massive databank – these are not simple systems.

Fully understanding the properties of OCP materials, and the manner in which these factors interact with the molecular biology of the cytoplasm and/or surface of cells, is vital for their successful translation to useful bioapplications. Literature reports confirm that the properties of OCPs vary widely and are related to composition, processing, and conditions in a complex way. Thus, the study of the basic properties must be conducted at a fundamental level by developing structure–property relationships. This, however, requires a greater understanding of the structure of OCPs both at the molecular and the supramolecular levels and the manner in which specific aspects of fabrication processes affect intermolecular interactions during the growth of the polymer chain. One fundamental, potentially confounding consideration toward achieving such understanding is the dynamic nature apparent in the transmission interplay(s) between OCP structure and material properties under varied conditions. It is inherently challenging to gain discrete information regarding a particular property (e.g., surface energy) without any influence on that particular property from another (e.g., surface roughness or dopant chemistry).

A further aspect that has not been described in great detail is the effect of service environments on the properties of OCPs. While the effects of aging on electrical properties have received most attention, the changes in mechanical and chemical properties are less well known. If OCPs are to be confidently used in service, their performance must be addressed at various temperatures, under varying strain rates and cyclic loads, in contact with various liquid or gaseous media and under the influence of applied electrical fields.

Within this context, recent studies have begun to explore the influence of changes in mechanical properties during redox cycling on the performance of PPy as an electromechanical actuator. Such studies highlight the problems, and also opportunities, that arise in these dynamic materials. The organization required to achieve the desired chemical and electrical properties will also determine the mechanical properties of any practical structure we are to make. All these three properties (chemical, electrical, and mechanical) are inextricably linked and individually capable of promoting specific medical bionic effects on cells that come into contact with the polymer surface.

References

1. Nobelprize.org. (2000) The Nobel Prize in Chemistry, *http://nobelprize.org/nobel_prizes/ chemistry/laureates/2000/* (accessed 25 November 2010).

2. Wallace, G.G., Spinks, G.M., Kane-Maguire, L.A.P., and Teasdale, P.R. (2008) *Organic Conducting Polymers: Intelligent Polymer Systems*, 3rd edn, CRC Press, Taylor & Francis Group, Boca Raton.

3. Hodgson, A.J., Gilmore, K., Small, C., Wallace, G.G., MacKenzie, I., Aoki, T., and Ogata, N. (1994) Reactive supramolecular assemblies of mucopolysaccharide, polypyrrole and protein as controllable biocomposites for a new generation of 'intelligent biomaterials'. *Supramol. Sci.*, **1**, 77–83.

4. Gilmore, K.J., Kita, M., Han, Y., Gelmi, A., Higgins, M.J., Moulton, S.E., Clark, G.M., Kapsa, R., and Wallace, G.G. (2009) Skeletal muscle cell proliferation and differentiation on polypyrrole substrates doped with extracellular matrix components. *Biomaterials*, **30**, 5292–5304.

5. Collier, J.H., Camp, J.P., Hudson, T.W., and Schmidt, C.E. (2000) Synthesis and characterization of polypyrrole-hyaluronic acid composite biomaterials for tissue engineering applications. *J. Biomed. Mater. Res.*, **50**, 574–584.

6. Yang, X., Too, C.O., Sparrow, L., Ramshaw, J., and Wallace, G.G. (2002) Polypyrrole-heparin system for the separation of thrombin. *React. Funct. Polym.*, **53**, 53–62.

7. Wu, Y.Z., Moulton, S.E., Too, C.O., Wallace, G.G., and Zhou, D.Z. (2004) Use of inherently conducting polymers and pulsed amperometry in flow injection analysis to detect oligonucleotides. *Analyst*, **129**, 585–588.

8. Misoska, V., Price, W.E., Ralph, S.F., Wallace, G.G., and Ogata, N. (2001) Synthesis, characterisation and ion transport studies on polypyrrole/deoxyribonucleic acid conducting polymer membranes. *Synth. Met.*, **123**, 279–286.

9. Campbell, T.E., Hodgson, A.J., and Wallace, G.G. (1999) Incorporation of erythrocytes into polypyrrole to form the basis of a biosensor to screen for Rhesus (D) blood groups and rhesus (D) antibodies. *Electroanal.*, **11**, 215–222.

10. Zhao, H.J. and Wallace, G.G. (1998) Polypyrrole poly(2-methoxyaniline-5-sulfonic acid) polymer composite. *Polym. Gels Networks*, **6**, 233–245.

11. Liu, X., Gilmore, K.J., Moulton, S.E., and Wallace, G.G. (2009) Electrical stimulation promotes nerve cell differentiation on polypyrrole/poly (2-methoxy-5 aniline sulfonic acid) composites. *J. Neural Eng.*, **6**, 065002.

12. Beck, F., Braun, P., and Oberst, M. (1987) Organic electrochemistry in the solid state-overoxidation of polypyrrole. Berichte der bunsen-gesellschaft-phys. Chem. *Chem. Phys.*, **91**, 967–974.

13. Freund, M., Bodalbhai, L., and Brajtertoth, A. (1991) Anion-excluding polypyrrole films. *Talanta*, **38**, 95–99.

14. Thompson, B.C., Moulton, S.E., Richardson, R.T., and Wallace, G.G. (2011) Effect of the anionic dopant on nerve growth and controlled release of a neurotrophic protein from polypyrrole. *Biomaterials*, **32**, 3822–3830.

15. Thompson, B.C., Moulton, S.E., Ding, J., Richardson, R., Cameron, A., O'Leary, S., Wallace, G.G., and Clark, G.M. (2006) Optimising the incorporation and release of a neurotrophic factor using conducting polypyrrole. *J. Control. Release*, **116**, 285–294.

16. Masdarolomoor, F., Innis, P.C., Ashraf, S., Kaner, R.B., and Wallace, G.G. (2006) Nanocomposites of polyaniline/poly(2-methoxyaniline-5-sulfonic acid). *Macromol. Rapid Commun.*, **27**, 1995–2000.

17. Reece, D.A., Innis, P.C., Ralph, S.F., and Wallace, G.G. (2002) Electrohydrodynamic synthesis of polypyrrole coated polyurethane colloidal dispersions using the electrocatalyst Tiron. *Colloids Surf., A*, **207**, 1–12.

18. Barisci, J.N., Hodgson, A.J., Liu, L., Wallace, G.G., and Harper, G. (1999) Electrochemical production of protein-containing polypyrrole colloids. *React. Funct. Polym.*, **39**, 269–275.

19. Davey, J.M., Innis, P.C., Ralph, S.E., Too, C.O., Wallace, G.G., and Partridge, A.C. (2000) Electrohydrodynamic synthesis, characterisation and metal uptake studies on polypyrrole colloids stabilised by polyvinylphosphate dopant. *Colloids Surf., A*, **175**, 291–301.

20. Aboutanos, V., Kane-Maguire, L.A.P., and Wallace, G.G. (2000) Electrosynthesis of polyurethane-based core-shell PAn center dot (+)-HCSA colloids. *Synth. Met.*, **114**, 313–320.

21. Cawdery, N., Obey, T.M., and Vincent, B. (1988) Colloidal dispersions of electrically conducting polypyrrole particles in various media. *J. Chem. Soc. Chem. Commun.*, 1189–1190.

22. Markham, G., Obey, T.M., and Vincent, B. (1990) The preparation and properties of dispersions of electrically-conducting polypyrrole particles. *Colloids Surf.*, **51**, 239–253.

23. Bjorklund, R.B. and Liedberg, B. (1986) Electrically conducting composites of colloidal polypyrrole and methylcellulose. *J. Chem. Soc. Chem. Commun.*, 1293–1295.

24. Jang, J., Oh, J.H., and Stucky, G.D. (2002) Fabrication of ultrafine conducting polymer and graphite nanoparticles. *Angew. Chem. Int. Ed.*, **41**, 4016–4019.

25. Yang, X.M., Dai, T.Y., Wei, M., and Lu, Y. (2006) Polymerization of pyrrole on a polyelectrolyte hollow-capsule microreactor. *Polymer*, **47**, 4596–4602.

26. Shi, W., Ge, D.T., Wang, J.X., Jiang, Z.Z., Ren, L., and Zhang, Q.Q. (2006) Heparin-controlled growth of polypyrrole nanowires. *Macromol. Rapid Commun.*, **27**, 926–930.

27. Qi, Z.G. and Lennox, R.B. (1997) in *Proceedings of the Third International Symposium on Microstructures and Microfabricated Systems* (eds P.J. Hesketh, G. Barna, and H.G. Hughes), Electrochemical Society Inc., Pennington, pp. 173–178.

28. Tan, S.N. and Ge, H.L. (1996) Investigation into vapour-phase

formation of polypyrrole. *Polymer,* **37**, 965–968.

29. Winther-Jensen, B., Chen, J., West, K., and Wallace, G. (2004) Vapor phase polymerization of pyrrole and thiophene using iron(III) sulfonates as oxidizing agents. *Macromolecules,* **37**, 5930–5935.

30. Winther-Jensen, B., Chen, J., West, K., and Wallace, G. (2005) 'Stuffed' conducting polymers. *Polymer,* **46**, 4664–4669.

31. Chen, J., Winther-Jensen, B., Lynam, C., Ngamna, O., Moulton, S., Zhang, W.M., and Wallace, G.G. (2006) A simple means to immobilize enzyme into conducting polymers via entrapment. *Electrochem. Solid-State Lett.,* **9**, H68–H70.

32. Thompson, B.C., Chen, J., Moulton, S.E., and Wallace, G.G. (2010) Nanostructured aligned CNT platforms enhance the controlled release of a neurotrophic protein from polypyrrole. *Nanoscale,* **2**, 499–501.

33. Krische, B. and Zagorska, M. (1989) The polythiophene paradox. *Synth. Met.,* **28**, C263–C268.

34. Sato, M., Tanaka, S., and Kaeriyama, K. (1986) Electrochemical preparation of conducting poly(3-Methylthiophene) – Comparison with polythiophene and poly(3-Ethylthiophene). *Synth. Met.,* **14**, 279–288.

35. Krische, B., Zagorska, M., and Hellberg, J. (1993) Bithiophenes as starting monomers for polythiophene syntheses. *Synth. Met.,* **58**, 295–307.

36. Krische, B. and Zagorska, M. (1989) Polythiophene synthesis by electropolymerization of thiophene and bithiophene. *Synth. Met.,* **33**, 257–267.

37. Yen, W., Chan, C.C., Jing, T., Jang, G.W., and Hsueh, K.F. (1991) Electrochemical polymerization of thiophenes in the presence of bithiophene or terthiophene – kinetics and mechanism of the

polymerization. *Chem. Mater.,* **3**, 888–897.

38. Faid, K., Cloutier, R., and Leclerc, M. (1993) Design of new electroactive polybithiophene derivatives. *Synth. Met.,* **55**, 1272–1277.

39. Xiao, Y.H., Li, C.M., Wang, S.Q., Shi, J.S., and Ooi, C.P. (2010) Incorporation of collagen in poly(3,4-ethylenedioxythiophene) for a bifunctional film with high bio- and electrochemical activity. *J. Biomed. Mater. Res. Part A,* **92A**, 766–772.

40. Asplund, M., Thaning, E., Lundberg, J., Sandberg-Nordqvist, A.C., Kostyszyn, B., Inganas, O., and von Holst, H. (2009) Toxicity evaluation of PEDOT/biomolecular composites intended for neural communication electrodes. *Biomed. Mater.,* **4**, 045009.

41. Zeng, H.J., Jiang, Y.D., Yu, J.S., and Xie, G.Z. (2008) Choline oxidase immobilized into conductive poly(3,4-ethylenedioxythiophene) film for choline detection. *Appl. Surf. Sci.,* **254**, 6337–6340.

42. Xiao, Y.H., Cui, X.Y., Hancock, J.M., Bouguettaya, M.B., Reynolds, J.R., and Martin, D.C. (2004) Electrochemical polymerization of poly(hydroxymethylated-3, 4-ethylenedioxythiophene) (PEDOT-MeOH) on multichannel neural probes. *Sens. Actuators, B Chem.,* **99**, 437–443.

43. Beck, F. and Barsch, U. (1993) The role of water in the electrodeposition and doping of polythiophene and 2 of its derivatives. *Macromol. Chem. Phys.,* **194**, 2725–2739.

44. Gningue-Sall, D., Fall, M., Dieng, M.M., Aaron, J.J., and Lacaze, P.C. (1999) Electrosynthesis and characterization of poly(3-methoxythiophene)-polybithiophene composite films prepared in micellar media on Pt and Fe substrates. *Phys. Chem. Chem. Phys.,* **1**, 1731–1734.

45. Fall, M., Dieng, M.M., Aaron, J.J., Aeiyach, S., and Lacaze, P.C. (2001)

Role of surfactants in the electrosynthesis and the electrochemical and spectroscopic characteristics of poly(3-methoxythiophene) films in aqueous micellar media. *Synth. Met.*, **118**, 149–155.

46. Bazzaoui, E.A., Aeiyach, S., and Lacaze, P.C. (1996) Electropolymerization of bithiophene on Pt and Fe electrodes in an aqueous sodium dodecylsulfate (SDS) micellar medium. *Synth. Met.*, **83**, 159–165.

47. Sakmeche, N., Aeiyach, S., Aaron, J.J., Jouini, M., Lacroix, J.C., and Lacaze, P.C. (1999) Improvement of the electrosynthesis and physicochemical properties of poly(3,4-ethylenedioxythiophene) using a sodium dodecyl sulfate micellar aqueous medium. *Langmuir*, **15**, 2566–2574.

48. Inoue, M.B., Velazquez, E.F., and Inoue, M. (1988) One-step chemical synthesis of doped polythiophene by use of copper(II) perchlorate as an oxidant. *Synth. Met.*, **24**, 223–229.

49. McCullough, R.D., Lowe, R.D., Jayaraman, M., Ewbank, P.C., Anderson, D.L., and Tristramnagle, S. (1993) Synthesis and physical-properties of regiochemically well-defined, head-to-tail coupled poly(3-Alkylthiophenes). *Synth. Met.*, **55**, 1198–1203.

50. Lock, J.P., Im, S.G., and Gleason, K.K. (2006) Oxidative chemical vapor deposition of electrically conducting poly(3,4-ethylenedioxythiophene) films. *Macromolecular*, **39**, 5326–5329.

51. Levermore, P.A., Chen, L.C., Wang, X.H., Das, R., and Bradley, D.D.C. (2007) Highly conductive poly(3,4-ethylenedioxythiophene) films by vapor phase polymerization for application in efficient organic light-emitting diodes. *Adv. Mater.*, **19** (17), 2379–2385.

52. Oh, S.G. and Im, S.S. (2002) Electroconductive polymer nanoparticles preparation and characterization of PANI and PEDOT nanoparticles. *Curr. Appl. Phys.*, **2**, 273–277.

53. Porter, T.L., Minore, D., and Sykes, A.G. (1995) Counterion and dopant-induced effects on the structure of electropolymerized polyaniline thin-films. *J. Vac. Sci. Technol., A Surf. Films*, **13**, 1286–1289.

54. Duic, L., Mandic, Z., and Kovacicek, F. (1994) The effect of supporting electrolyte on the electrochemical synthesis, morphology, and conductivity of polyaniline. *J. Polym. Sci., Part A: Polym. Chem.*, **32**, 105–111.

55. Yasuda, A. and Shimidzu, T. (1993) Chemical and electrochemical analyses of polyaniline prepared with FeCl$_3$. *Synth. Met.*, **61**, 239–245.

56. Kim, B.J., Oh, S.G., Han, M.G., and Im, S.S. (2001) Synthesis and characterization of polyaniline nanoparticles in SDS micellar solutions. *Synth. Met.*, **122**, 297–304.

57. Moulton, S.E., Innis, P.C., Kane-Maguire, L.A.P., Ngamna, O., and Wallace, G.G. (2004) Polymerisation and characterisation of conducting polyaniline nanoparticle dispersions. *Curr. Appl. Phys.*, **4**, 402–406.

58. Huang, J.X., Virji, S., Weiller, B.H., and Kaner, R.B. (2003) Polyaniline nanofibers: facile synthesis and chemical sensors. *J. Am. Chem. Soc.*, **125**, 314–315.

59. Huang, J.X. and Kaner, R.B. (2004) A general chemical route to polyaniline nanofibers. *J. Am. Chem. Soc.*, **126**, 851–855.

60. Li, D. and Kaner, R.B. (2006) Shape and aggregation control of nanoparticles: Not shaken, not stirred. *J. Am. Chem. Soc.*, **128**, 968–975.

61. Tanaka, K., Shichiri, T., Wang, S.G., and Yamabe, T. (1988) A study of the electropolymerization of thiophene. *Synth. Met.*, **24**, 203–215.

62. Tourillon, G. and Garnier, F. (1983) Effect of dopant on the physicochemical and electrical-properties

of organic conducting polymers. *J. Phys. Chem.*, **87**, 2289–2292.

63. Tourillon, G. and Garnier, F. (1984) Electrochemical doping of polythiophene in aqueous-medium – electrical-properties and stability. *J. Electroanal. Chem.*, **161**, 407–414.

64. Bredas, J.L., Chance, R.R., and Silbey, R. (1982) Comparative theoretical-study of the doping of conjugated polymers – polarons in polyacetylene and polyparaphenylene. *Phys. Rev. B*, **26**, 5843–5854.

65. Chen, J., Too, C.O., Wallace, G.G., and Swiegers, G.F. (2004) Redox-active conducting polymers incorporating ferrocenes 2. Preparation and characterisation of polypyrroles containing propyl- and butyl-tethered[1.1]ferrocenophane. *Electrochim. Acta*, **49**, 691–702.

66. Andrieux, C.P., Audebert, P., Hapiot, P., Nechtschein, M., and Odin, C. (1991) Fast scan rate cyclic voltammetry for conducting polymers electropolymerized on ultramicroelectrodes. *J. Electroanal. Chem.*, **305**, 153–162.

67. Marque, P., Roncali, J., and Garnier, F. (1987) Electrolyte effect on the electrochemical properties of poly(3-Methylthiophene) thin-films. *J. Electroanal. Chem.*, **218**, 107–118.

68. Chen, S.A. and Tsai, C.C. (1993) Structure properties of conjugated conductive polymers.2. 3-Ether-substituted polythiophenes and poly(4-Methylthiophene)S. *Macromolecules*, **26**, 2234–2239.

69. Breukers, R.D., Gilmore, K.J., Kita, M., Wagner, K.K., Higgins, M.J., Moulton, S.E., Clark, G.M., Officer, D.L., Kapsa, R.M.I., and Wallace, G.G. (2010) Creating conductive structures for cell growth: growth and alignment of myogenic cell types on polythiophenes. *J. Biomed. Mater. Res. Part A* **95A**, 256–268.

70. Grzeszczuk, M. and Zabinskaolszak, G. (1993) Ionic transport in polyaniline film electrodes – an impedance study. *J. Electroanal. Chem.*, **359**, 161–174.

71. Lacroix, J.C. and Diaz, A.F. (1988) Electrolyte effects on the switching reaction of polyaniline. *J. Electrochem. Soc.*, **135**, 1457–1463.

72. Foot, P.J.S. and Simon, R. (1989) Electrochromic properties of conducting polyanilines. *J. Phys. D: Appl. Phys.*, **22**, 1598–1603.

73. Bartlett, P.N. and Wallace, E.N.K. (2000) The oxidation of beta-nicotinamide adenine dinucleotide (NADH) at poly(aniline)-coated electrodes Part II. Kinetics of reaction at poly(aniline)- poly(styrenesulfonate) composites. *J. Electroanal. Chem.*, **486**, 23–31.

74. Keselowsky, B.G., Collard, D.M., and Garcia, A.J. (2004) Surface chemistry modulates focal adhesion composition and signaling through changes in integrin binding. *Biomaterials*, **25**, 5947–5954.

75. Kiddie, G., McLean, D., Van Ooyen, A., and Graham, B. (2005) in *Development, Dynamics and Pathology of Neuronal Networks: from Molecules to Functional Circuits*, Progress in Brain Research, Vol. **147** (eds M. Kamermans, C.N. Levelt, A. Van Ooyen, G.J.A. Ramakers, and P.R. Roelfsema), Elsevier, Amsterdam, pp. 67–80.

76. John, R. and Wallace, G.G. (1993) Doping dedoping of polypyrrole – a study using current-measuring and resistance-measuring techniques. *J. Electroanal. Chem.*, **354**, 145–160.

77. Ge, H. and Wallace, G.G. (1992) Ion-exchange properties of polypyrrole. *React. Polym.*, **18**, 133–140.

78. Sadik, O.A. and Wallace, G.G. (1993) Effect of polymer composition on the detection of electroinactive species using conductive polymers. *Electroanalytical*, **5**, 555–563.

79. Shimidzu, T., Ohtani, A., and Honda, K. (1988)

Charge-controllable polypyr-role poly-electrolyte composite membranes.3. Electrochemical deionization system constructed by anion-exchangeable and cation-exchangeable polypyrrole electrodes. *J. Electroanal. Chem.*, **251**, 323–337.

80. Zhong, C.J. and Doblhofer, K. (1990) Polypyrrole-based electrode coatings switchable electrochemi-cally between the anion-exchanger and cation-exchanger states. *Elec-trochim. Acta*, **35**, 1971–1976.

81. Naoi, K., Lien, M., and Smyrl, W.H. (1991) Quartz crystal microbalance study – ionic motion across conducting polymers. *J. Electrochem. Soc.*, **138**, 440–445.

82. Lewis, T.W., Wallace, G.G., and Smyth, M.R. (1999) Electrofunc-tional polymers: their role in the development of new analytical systems. *Analyst*, **124**, 213–219.

83. Wallace, G.G. and Kane-Maguire, L.A.P. (2002) Manipulating and monitoring biomolecular interactions with conducting elec-troactive polymers. *Adv. Mater.*, **14**, 953–958.

84. Wallace, G.G., Maxwell, K.E., Lewis, T.W., Hodgson, A.J., and Spencer, M.J. (1990) New conducting polymer affinity-chromatography station-ary phases. *J. Liq. Chromatogr.*, **13**, 3091–3110.

85. Barnett, D., Laing, D.G., Skopec, S., Sadik, O., and Wallace, G.G. (1994) Determination of P-cresol (and Other Phenolics) using a conducting polymer-based electro-immunological sensing system. *Anal. Lett.*, **27**, 2417–2429.

86. Sadik, O.A. and Van Emon, J.M. (1996) Applications of electro-chemical immunosensors to environmental monitoring. *Biosens. Bioelectron.*, **11**, I–XI.

87. Bender, S. and Sadik, O.A. (1998) Direct electrochemical immunosen-sor for polychlorinated biphenyls. *Environ. Sci. Technol.*, **32**, 788–797.

88. Foulds, N.C. and Lowe, C.R. (1988) Immobilization of glucose-oxidase in ferrocene-modified pyrrole poly-mers. *Anal. Chem.*, **60**, 2473–2478.

89. Couves, L.D. and Porter, S.J. (1989) Polypyrrole as a potentiometric glucose sensor. *Synth. Met.*, **28**, C761–C768.

90. Adeloju, S.B., Shaw, S.J., and Wallace, G.G. (1997) Pulsed-amperometric detection of urea in blood samples on a conducting polypyrrole-urease biosensor. *Anal. Chim. Acta*, **341**, 155–160.

91. Compagnone, D., Federici, G., and Bannister, J.V. (1995) A new conducting polymer glucose sen-sor based on polythianaphthene. *Electroanalytical*, **7**, 1151–1155.

92. Lu, W., Zhou, D.Z., and Wallace, G.G. (1998) Enzymatic sensor based on conducting polymer coat-ings on metallised membranes. *Anal. Commun.*, **35**, 245–248.

93. Stevenson, G., Moulton, S.E., Innis, P.C., and Wallace, G.G. (2010) Polyterthiophene as an electros-timulated controlled drug release material of therapeutic levels of dexamethasone. *Synth. Met.*, **160**, 1107–1114.

94. Moulton, S.E., Imisides, M.D., Shepherd, R.L., and Wallace, G.G. (2008) Galvanic coupling conduct-ing polymers to biodegradable Mg initiates autonomously powered drug release. *J. Mater. Chem.*, **18**, 3608–3613.

95. Richardson, R.T., Thompson, B., Moulton, S., Newbold, C., Lum, M.G., Cameron, A., Wallace, G., Kapsa, R., Clark, G., and O'Leary, S. (2007) The effect of polypyrrole with incorporated neurotrophin-3 on the promotion of neurite out-growth from auditory neurons. *Biomaterials*, **28**, 513–523.

96. Thompson, B.C., Richardson, R.T., Moulton, S.E., Evans, A.J., O'Leary, S., Clark, G.M., and Wallace, G.G. (2010) Conducting polymers, dual neurotrophins and pulsed electri-cal stimulation – dramatic effects

on neurite outgrowth. *J. Control. Release*, **141**, 161–167.

97. Wang, J. and Jiang, M. (2000) Toward genolelectronics: nucleic acid doped conducting polymers. *Langmuir*, **16**, 2269–2274.

98. Sadik, O.A. and Wallace, G.G. (1993) Pulsed amperometric detection of proteins using antibody containing conducting polymers. *Anal. Chim. Acta*, **279**, 209–212.

99. Barnett, D., Laing, D.G., Skopec, S., Sadik, O., and Wallace, G.G. (1994) Determination of P-Cresol (and Other Phenolics) using a conducting polymer-based electro-immunological sensing system. *Anal. Lett.*, **27**, 2417–2429.

100. Gooding, J.J., Wasiowych, C., Barnett, D., Hibbert, D.B., Barisci, J.N., and Wallace, G.G. (2004) Electrochemical modulation of antigen-antibody binding. *Biosens. Bioelectron.*, **20**, 260–268.

101. Kotkar, D., Joshi, V., and Ghosh, P.K. (1988) Towards chiral metals – synthesis of chiral conducting polymers from optically-active thiophene and pyrrole derivatives. *J. Chem. Soc. Chem. Commun.*, 917–918.

102. Rughooputh, S., Nowak, M., Hotta, S., Heeger, A.J., and Wudl, F. (1987) Soluble conducting polymers – the poly(3-Alkylthienylenes). *Synth. Met.*, **21**, 41–50.

103. Andersson, M., Ekeblad, P.O., Hjertberg, T., Wennerstrom, O., and Inganas, O. (1991) Polythiophene with a free amino-acid side-chain. *Polym. Commun.*, **32**, 546–548.

104. Ikenoue, Y., Tomozawa, H., Saida, Y., Kira, M., and Yashima, H. (1991) Evaluation of electrochromic fast-switching behavior of self-doped conducting polymer. *Synth. Met.*, **40**, 333–340.

105. Patil, A.O., Ikenoue, Y., Basescu, N., Colaneri, N., Chen, J., Wudl, F., and Heeger, A.J. (1987) Self-doped conducting polymers. *Synth. Met.*, **20**, 151–159.

106. Bryce, M.R., Chissel, A., Kathirgamanathan, P., Parker, D., and Smith, N.R.M. (1987) Soluble, conducting polymers from 3-substituted thiophenes and pyrroles. *J. Chem. Soc. Chem. Commun.*, 466–467.

107. Torres, W. and Fox, M.A. (1990) Poly(N-3-Thenylphthalimide) – conductivity and spectral properties. *Chem. Mater.*, **2**, 158–162.

108. Syed, A.A. and Dinesan, M.K. (1990) Polyaniline – reaction stoichiometry and use as an ion-exchange polymer and acid-base indicator. *Synth. Met.*, **36**, 209–215.

109. Syed, A.A. and Dinesan, M.K. (1992) Poly(Aniline) – a conducting polymer as a novel anion-exchange resin. *Analyst*, **117**, 61–66.

110. Teasdale, P.R. and Wallace, G.G. (1994) Characterizing the chemical interactions that occur on polyaniline with inverse thin-layer chromatography. *Polym. Int.*, **35**, 197–205.

111. Shinohara, H., Chiba, T., and Aizawa, M. (1988) Enzyme microsensor for glucose with an electro-chemically synthesized enzyme polyaniline film. *Sens. Actuators*, **13**, 79–86.

112. Tatsuma, T., Ogawa, T., Sato, R., and Oyama, R. (2001) Peroxidase-incorporated sulfonated polyaniline-polycation complexes for electrochemical sensing of H_2O_2. *J. Electroanal. Chem.*, **501**, 180–185.

113. Nagarajan, R., Liu, W., Kumar, J., Tripathy, S.K., Bruno, F.F., and Samuelson, L.A. (2001) Manipulating DNA conformation using intertwined conducting polymer chains. *Macromolecules*, **34**, 3921–3927.

114. Sun, B., Jones, J.J., Burford, R.P., and Skyllas-Kazacos, M. (1989) Stability and mechanical-properties of electrochemically prepared conducting polypyrrole films. *J. Mater. Sci.*, **24**, 4024–4029.

115. Wang, X.S., Xut, J.K., Shi, G.Q., and Lu, X. (2002) Microstructure-mechanical properties relationship in conducting polypyrrole films. *J. Mater. Sci.*, **37**, 5171–5176.

116. Wynne, K.J. and Street, G.B. (1985) Poly(Pyrrol-2-Ylium Tosylate) – electrochemical synthesis and physical and mechanical-properties. *Macromolecules*, **18**, 2361–2368.

117. Moss, B.K., Burford, R.P., and Skyllas-Kazacos, M. (1989) Electrically Conductive Polypyrroles. *Mater. Forum*, **13**, 35–42.

118. Buckley, L.J., Roylance, D.K., and Wnek, G.E. (1987) Influence of dopant ion and synthesis variables on mechanical-properties of polypyrrole films. *J. Polym. Sci. Part B Polym. Phys.*, **25**, 2179–2188.

119. Cvetko, B.F., Brungs, M.P., Burford, R.P., and Skyllas-Kazacos, M. (1987) Conductivity measurements of electrodeposited polypyrrole. *J. Appl. Electrochem.*, **17**, 1198–1202.

120. Gandhi, M., Spinks, G.M., Burford, R.P., and Wallace, G.G. (1995) Film substructure and mechanical-properties of electrochemically prepared polypyrrole. *Polymer*, **36**, 4761–4765.

121. Gandhi, M.R., Murray, P., Spinks, G.M., and Wallace, G.G. (1995) Mechanism of electromechanical actuation in polypyrrole. *Synth. Met.*, **73**, 247–256.

122. Roncali, J. (1992) Conjugated poly(Thiophenes) – synthesis, functionalization, and applications. *Chem. Rev.*, **92**, 711–738.

123. Jin, S., Cong, S.X., Xue, G., Xiong, H.M., Mansdorf, B., and Cheng, S.Z.D. (2002) Anisotropic polythiophene films with high conductivity and good mechanical properties via a new electrochemical synthesis. *Adv. Mater.*, **14**, 1492–1496.

124. Yoshino, K., Tabata, M., Satoh, M., Kaneto, K., and Hasegawa, T. (1985) *Tech. Rep.*, **35**, 231.

125. Ito, M., Tsuruno, A., Osawa, S., and Tanaka, K. (1988) Morphology and mechanical-properties of electrochemically prepared polythiophene films. *Polymer*, **29**, 1161–1165.

126. Fedorko, P., Fraysse, J., Dufresne, A., Planes, J., Travers, J.P., Olinga, T., Kramer, C., Rannou, P., and Pron, A. (2001) New counterion-plasticized polyaniline with improved mechanical and thermal properties: comparison with PANI-CSA. *Synth. Met.*, **119**, 445–446.

127. Dufour, B., Rannou, P., Fedorko, P., Djurado, D., Travers, J.P., and Pron, A. (2001) Effect of plasticizing dopants on spectroscopic properties, supramolecular structure, and electrical transport in metallic polyaniline. *Chem. Mater.*, **13**, 4032–4040.

128. Li, J., Wang, E., Green, M., and West, P.E. (1995) In-situ AFM study of the surface-morphology of polypyrrole film. *Synth. Met.*, **74**, 127–131.

129. Barisci, J.N., Stella, R., Spinks, G.M., and Wallace, G.G. (2000) Characterisation of the topography and surface potential of electrodeposited conducting polymer films using atomic force and electric force microscopies. *Electrochim. Acta*, **46**, 519–531.

130. Silk, T., Hong, Q., Tamm, J., and Compton, R.G. (1998) AFM studies of polypyrrole film surface morphology – I. The influence of film thickness and dopant nature. *Synth. Met.*, **93**, 59–64.

131. Silk, T., Hong, Q., Tamm, J., and Compton, R.G. (1998) AFM studies of polypyrrole film surface morphology – II. Roughness characterization by the fractal dimension analysis. *Synth. Met.*, **93**, 65–71.

132. Yoon, C.O., Sung, H.K., Kim, J.H., Barsoukov, E., Kim, J.H., and Lee, H. (1999) The effect of low-temperature conditions on the electrochemical polymerization of

polypyrrole films with high density, high electrical conductivity and high stability. *Synth. Met.*, **99**, 201–212.

133. Johansson, F., Carlberg, P., Danielsen, N., Montelius, L., and Kanje, M. (2006) Axonal outgrowth on nano-imprinted patterns.

Biomaterials, **27**, 1251–1258.

134. Li, N.Z. and Folch, A. (2005) Integration of topographical and biochemical cues by axons during growth on microfabricated 3-D substrates. *Exp. Cell Res.*, **311**, 307–316.

4
Organic Conductors – Biological Applications

Essentially, medical bionic devices can be subclassified into three main application categories, as discussed below:

- **Functional medical bionics:** Pacemakers, cochlear implants, and deep brain stimulators function via delivery of electrical stimulation to excitable cell systems such as muscle, the peripheral nervous system, and the central nervous system (CNS), respectively. In such applications, the organic conductor component is required to restore a function that has been lost through some sort of tissue trauma or through some type of hereditary or acquired disease syndrome. The requirements for the electrode materials used in such devices include long-term maintenance of conductivity, long-term functional and materials integrity, and minimal support of introduced microbial infection.
- **Integrated sensory and motor bionics:** Considerable interest in the design of prosthetics that are able to "feel" has resulted in the necessity for such proposed medical bionic devices to integrate with tissues; one such application involves the integration of robotic devices with the nervous system [1]. The electrode materials requirements are similar to those for functional medical bionics.
- **Regenerative and protective bionics:** Involves the promotion of regenerative or trophic activity in cell systems that collectively compose a particular tissue. This can be achieved either by stimulating specific cells to trigger and facilitate a regenerative or cytoprotective response [2] or by providing a facilitatory matrix on which tissue can regenerate [3]. In addition to the electrode materials requirements outlined above, the electrodes used in regenerative devices would preferably be biodegradable, as would all the materials that comprise the device.

4.1
Carbon Structures for Medical Bionics

Over the past two decades, numerous strategies orientated toward the utilization of specific chemical and physical properties of carbon materials

Organic Bionics, First Edition. Gordon G. Wallace, Simon E. Moulton, Robert M.I. Kapsa,
and Michael J. Higgins.
© 2012 Wiley-VCH Verlag GmbH & Co. KGaA. Published 2012 by Wiley-VCH Verlag GmbH & Co. KGaA.

for biomedical applications have been documented. As detailed in the Chapter 2 on Carbon, section 2.1, carbon occurs in different lattice forms; from amorphous through to intermediate to crystalline (e.g., hexagonal or regular lattice diamond). In addition, carbon materials are able to be fashioned into highly effective medical device configurations through a wide spectrum of fabrication methods [4]. Carbon is an ideal biomaterial and has been used for a long time as a construction element; initially in the form of highly oriented pyrolytic graphite (HOPG) and composites containing carbon fibers [5]. These groups of "medically relevant" carbons elicit negligible (if any) adverse reaction in the host tissue and, conversely, are relatively unaffected by the host tissue environment [6] into which they are introduced.

4.1.1
Carbon-Based Electrodes for Medical Bionics

Pyrolytic carbon technology has been applied to cardiac pacing electrodes since 1982 [7]. In cardiac pacemakers, carbon has been shown to enhance the efficiency of energy transfer from the pacing tip to the adjacent myocardium, with negligible levels of thrombogenicity, low impedance, and negligible polarization [8, 9]. As a direct result of these characteristics, activated pyrolytic carbon was incorporated into the very first ventricular lead (SlOO; Sorin Biomedica Cardio SpA., Saluggia, Italy), whereby remarkably low defibrillation threshold values were achieved and maintained over time even in chronic conditions [10, 11].

The pioneering work of the 1980s, combined with significant advances in carbon synthesis protocols, has resulted in the introduction of low-polarization carbon electrodes and subsequent development of highly effective, light-weight (<25 g) new-generation pacemakers. In 1992, the SlOO/4 ventricular passive fixation lead was produced with a silicone elastomer insulating sheath, and a hemispherical electrode surface area that was further reduced in size to $4\,mm^2$. In this version of the original S1OO lead, the stimulating tip was made of a graphite core coated by a thin (100–150 μm) activated pyrolytic carbon layer. In 1995, Alt *et al.* [12] demonstrated significantly improved defibrillation thresholds with a new epicardial carbon electrode compared with a standard epicardial titanium patch.

Electrical stimulation of nerve tissue has now become widely employed in the treatment of acquired or hereditary neural defects by prosthetic neural implants such as the Cochlear electrode [13]. In addition, deep brain stimulation electrode protocols have emerged for neutralizing aberrant neural activity in the CNS for the treatment of movement disorders [14] and, more recently, for treatment-resistant neuropsychiatric illness [15]. Each of these applications involves an implanted electrode or electrode array that stimulates target neurons to modulate their function. This presents an opportunity for the use of carbon-based technology to improve efficacy of such electrodes. The rationale demonstrated by the vast improvements achieved through carbon coating of pacing electrodes has been widely assimilated toward developing carbon as an effective neurostimulatory or sensory electrode material [16–18].

Filamentous carbon fiber is an extremely biocompatible material that does not corrode and elicits almost no foreign body response. In addition, these fibers are able to be produced in a variety of shapes and configurations from "brush" to "needle" formats and efficiently deliver electrical stimulation *in vivo*. Carbon fiber electrodes can be readily produced with smaller geometric dimensions than metal electrodes, and their lower impedance provides better signal-to-noise resolution. This design has been further improved over the years by introducing new methods by which to etch the carbon fiber tip into desired geometries [19, 20] and speed up the manufacturing process [21, 22] (Figure 4.1).

Owing to their excellent conductive characteristics, carbon fiber electrodes are increasingly finding application in high-resolution recording of neural

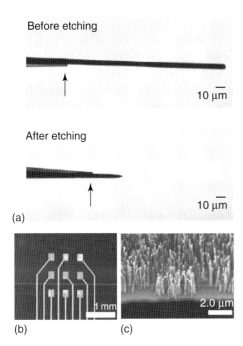

Figure 4.1 Photomicrographs of multi-barrel electrode tip (a) before (top) and after (bottom) etching the carbon fiber. Arrows indicate edge of glass. (Adapted with permission from Millar and Pelling [22].) Scanning electron microscopy (SEM) micrograph of 3 × 3 electrode array (b), as grown vertically aligned carbon nanoforest on electrodes (c). (Adapted with permission from de Asis and coworkers [23].)

(CNS) activity [24] and for the stimulation of bone and/or soft tissue growth [25]. In terms of the latter (bone) application, carbon-based electrodes are being investigated as suitable to be used in the electrical bone growth stimulator (EBGS). When used as a cathode material in conjunction with an implantable constant direct current device, carbon fiber was shown to be an effective electrode material for the stimulation of bone and fibrous tissue within the medullary canal of rabbit tibiae [25].

4.2
Carbon Nanotubes

Carbon nanotubes (CNTs) possess exceptional electrochemical, mechanical, and chemical properties by which they lend themselves to medical bionic applications. For example, Young's modulus of a CNT exceeds 1 TPa, about five times stronger than steel [26], yet CNTs can be bent and twisted to large angles without breaking [27]. This combination of strength, flexibility, and electrical conductivity is highly desirable for the development of medical bionic electrodes. The unique electrical and mechanical properties of CNTs make them excellent candidates for neural interfaces, but their adoption hinges on finding approaches for "humanizing" their composites. There has been recent progress in the use of CNTs in preservation of cells, delivery of growth factors or genes, and as scaffolding to promote integration with host tissue. CNTs can also be incorporated into other scaffold materials to provide structural reinforcement or confer novel properties, such as electrical conductivity, which may aid in directing cell growth [28]. The biocompatibility of CNTs for implantable electrodes has been extensively studied using a wide range of cell types such as nerve [16–18], endothelial [29], and stem cells [30–32]. Considerable ambiguity within the relevant literature has necessitated careful consideration of the context in which CNTs are able to be used in proposed medical bionics devices.

4.2.1
Neural Applications

The earliest report of the growth of rat hippocampal neurons on CNTs [33] described containment of neurite branching on a multiwalled nanotube (MWNT) surface compared to standard growth substratum. This important early result gave rise to a significant body of literature that has highlighted the versatility of functionalized or nonfunctionalized CNT formats (i.e., single-walled nanotubes (SWNTs, MWNTs) as novel neuromodulatory substrata for neural cell growth. Applications for the use of CNTs in tissues of the nervous system now encompass stimulation as well as recording of neural function [34]. For instance, Mazzatenta *et al.* [34] developed an integrated

SWNT–neuron system to test whether electrical stimulation delivered via SWNT can induce neuronal signaling. They showed that hippocampal neurons grown on pure SWNT were viable and that the electrical interactions occurring in SWNT–neuron hybrid systems clearly indicate that SWNTs can directly stimulate brain circuit activity (Figure 4.2). The unique properties of CNTs and the application of nanotechnology to the nervous system may have a tremendous impact on the future developments of microsystems for neural prosthetics as well as immediate benefits for medical bionic research. Electrical stimulation of nerve cells is widely employed in neural prostheses (for hearing, [35] vision [36], and potentially for limb movement restoration [37]), in clinical therapies (e.g., Parkinson's disease, dystonia, chronic pain [14]), and in basic neuroscience studies. In all of these applications, an implanted microelectrode or microelectrode array stimulates neurons and modulates their behavior.

An ideal stimulating array needs to be both efficacious and safe [38]. However, functional stimulation (especially in the CNS) often demands small electrodes with high current density, which conflicts with electrochemical safety requirements. Undesirable electrochemical reactions not only damage the electrodes but also cause abnormalities in neural function and cell structure [39]. To enhance electrode sensitivity or increase electric charge for stimulation, the impedance must be minimized [40]. The biocompatibility of SWNTs has been shown previously [33, 41, 42], and neuronal adhesion, survival, and growth can be modulated by using appropriate SWNT–polymer

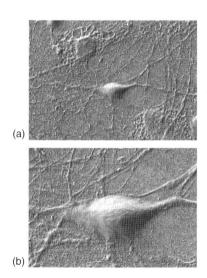

(a)

(b)

Figure 4.2 Scanning electron microscopy images of (a) cultured hippocampal neurons on SWNTs. High magnification (b) shows neurons grown on SWNTs (10 days). (Adapted with permission from Mazzatenta and coworkers [34].)

conjugates [43]. For example, Lovat *et al.* [42] recently demonstrated that purified CNTs provide suitable growth surfaces for neurons and promote an increase in the efficacy of neural signal transmission. Recent reports indicate the possibility of stimulating isolated neurons in culture via SWNT [44]. Nanofibers have been reported to minimize astrocyte reactions, thereby leading to reduction in glial scar tissue [45] associated with implantation. In addition, the nanoscale dimension of SWNTs allows molecular interactions with neurons [46] that will determine the electrical interfacing of SWNTs to neural cells.

The outstanding mechanical strength of CNTs can also be utilized for structural reinforcement of medical bionic interfaces. Beyond randomly oriented blends of nanotubes and polymers, CNTs can be arranged into ordered biomaterials, which can provide mechanical support against *in vivo* forces, such that the predefined 3D structure is maintained during tissue development. L929 mouse fibroblasts have been grown on such scaffolds [47]. Likewise, neurons can be grown on patterned arrays of CNTs to create ordered neural networks [48], and CNTs can be electrospun with biopolymers to form aligned matrices [49].

The intrinsically large surface areas ($700–1000\,\mathrm{m^2\,g^{-1}}$), extremely high conductance, and aspect ratios of CNTs provide rationale for their use as surface coatings for electrodes. Their properties confer the potential to reduce impedance and increase charge transfer compared to conventional microelectrodes. Keefer *et al.* [50] have established techniques for coating metal electrodes with CNTs and tested their function in cultured neuronal networks, the motor cortex of anaesthetized rats, and area V4 of rhesus macaques performing a visual task. Their results indeed showed that CNT-modified electrodes are robust, have greatly decreased impedances, lower susceptibility to noise, and increased ability to activate neurons when used for electrical stimulation, when compared to conventional metal electrodes.

4.2.2
Muscle Regeneration

MacDonald *et al.* [51] prepared composite materials composed of a collagen matrix with embedded CNTs by mixing solubilized collagen with solutions of carboxylated SWNTs. Living smooth muscle cells were incorporated at the collagen gelation stage to produce cell-seeded collagen-CNT composite matrices. They concluded that such collagen-CNT composite matrices may be useful as scaffolds in tissue regeneration. Garibaldi *et al.* [52] conducted *in vitro* studies to explore the cytocompatibility of purified carbon nanofibers with cardiomyocytes. Their findings showed that CNTs possess no evident short-term toxicity and can be considered biocompatible with cultured cardiomyocytes.

4.2.3
Bone

It is expected that CNTs can be applied to biomaterials used for bone [53]. Biomaterials used in the treatment of fractures or arthroplasty must possess high strength, and CNTs are known to improve this material property when used as a component of polymer composites. In addition, the ability of CNTs to impart electrical conductivity to these compounds is attractive in terms of stimulated growth of osteoblast cells and subsequent bone tissue.

Yadav *et al.* [54] explored the use of MWNTs as reinforcing agents to strengthen hydroxyapatite–gelatin (Hap–Gel) nanocomposites for artificial bone grafting applications without significantly compromising biocompatibility. Hap–Gel composites, reinforced with various proportions of MWNTs, were synthesized to optimize the MWNT content in the composites, yielding commendable improvement in their strength. Morphological studies reveal that the MWNTs act as templates for nucleation of Hap crystals. The biocompatibility of MWNT-reinforced Hap–Gel composites were evaluated in an animal model through the histopathological investigation of tissues from skin, kidney, and liver. On examination, no noticeable alteration due to toxicity was found for lower concentration of MWNTs (Figure 4.3). Mild reversible changes in the liver and tubular damage in the kidney have been observed for higher concentrations (4 wt% of MWNTs). It can be inferred from the findings that MWNTs, in proportions <4%, can successfully be used to reinforce the Hap–Gel nanocomposite to improve its mechanical properties. However, how safe these CNT-reinforced bone implants would be when used for prolonged periods in actual physiological conditions needs to be investigated further.

During the early stages of bone formation, collagen serves as a nucleation site for the deposition of hydroxyapatite, the principle component of bone. Phosphate-substituted CNTs can substitute for collagen to direct the crystallization of hydroxyapatite, reaching a thickness of 3 mm after 14 days of mineralization [55]. The high aspect ratio of CNTs should allow the scaffold to be aligned and to more closely mimic bone *in vivo*

This CNT blend approach has also been investigated by Supronowicz *et al.* [56], who investigated the effect of culturing osteoblasts on the surface of polylactic acid/CNT nanocomposites followed by exposure to electric stimulation (10 mA at 10 Hz) for 6 h per day for various periods of time. They observed a 46% increase in cell proliferation after 2 days, a 307% increase in the concentration of extracellular calcium after 21 consecutive days, and upregulation of mRNA expression for collagen type-I after both 1 and 21 consecutive days. These results provide evidence that electrical stimulation delivered through novel, current-conducting polymer/nanophase composites promotes osteoblast functions that are responsible for the chemical composition of the organic and inorganic phases of bone. Furthermore,

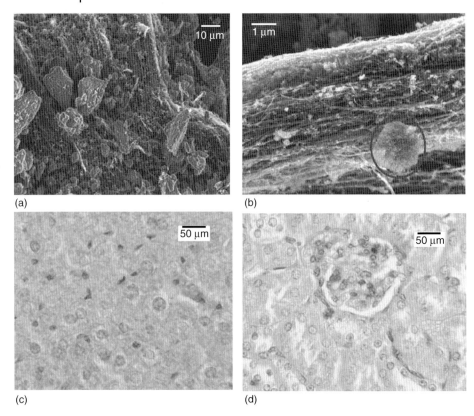

Figure 4.3 SEM micrograph of the composite with (a) 4 wt% MWNT at lower magnification showing overall morphology; (b) same at higher magnification showing an Hap particle (O circled) adhered to a bundle of MWNTs. Histological photomicrographs of control tissues from (c) liver; (d) kidney exposed to composites having 1% reinforcements, showing no significant change due to toxicity. (Adapted with permission from Yadav and coworkers [54].)

this evidence highlights important aspects of the cellular/molecular-level mechanisms involved in new bone formation under electrical stimulation.

Zanello *et al.* [57] also explored the use of CNTs as suitable scaffold materials for osteoblast proliferation and bone formation. With the aim of controlling cell growth, rat osteosarcoma 17/2.8 cells were cultured on chemically modified SWNTs and MWNTs. CNTs carrying neutral electric charge sustained the highest cell growth and production of plate-shaped crystals. There was a dramatic change in cell morphology in osteoblasts cultured on MWNTs, which correlated with changes in plasma membrane function.

4.2.4
Stem Cells

One of the key challenges to engineering neural interfaces is to minimize the extent of inflammatory response toward implanted electrodes. One potential approach is to manufacture materials that conform better with structural aspects of living tissues and, possibly, by utilizing stem cells, to generate a more compliant electrode/neural interface. More recently, the interaction between CNTs and stem cells has been investigated [30–32]. Collectively, the unique biofacilitative properties of CNTs combined with the broad versatility of human mesenchymal stem cells (hMSCs) provide exciting prospects for novel therapeutic modalities. However, little is known about the impact of CNTs on mesenchymal stem cells (MSCs) behavior, and in light of CNT teratogenicity issues reported recently, it is of critical importance that CNT effects on MSC growth and differentiation are more fully understood before they are used in any human application. Mooney *et al.* [32] reported no observable negative effect from CNTs on hMSC renewal, metabolic activity, and differentiation. Furthermore, they tracked the intracellular movement of CNT through the cytoplasm to a nuclear location and assessed effects on cellular ultrastructure. McCullen *et al.* [58] fabricated electrospun nanocomposite scaffolds by encapsulating MWNTs in poly-L-lactic acid (PLA) nanofibers in which the MWNTs were aligned along the axis of the fibers. Adipose-derived hMSCs were seeded onto PLA-containing MWNTs and were shown to survive and proliferate on the PLA-MWNT composites in a closely packed and longitudinally aligned manner.

The influence of carboxylic-functionalized SWNTs on cell adhesion, spreading, and cell lineage commitment of hMSCs has been evaluated [59]. hMSCs were cultured on a thin meshlike layer of SWNTs with a vertical height <100 nm with the influence of the SWNT film being significant on the cell spreading and focal adhesion distribution. Cells spread better on an SWNT film than on an uncoated cover slip (control), resulting in larger cell area and significantly greater filopodial extension at the cell boundaries. Cytoskeleton arrangements were observed to be less oriented in the cells cultured on an SWNT film as compared to control. Neurogenic markers such as nestin, glial fibrillary acidic protein, and microtubule-associated protein 2 genes were transiently upregulated (specific mRNA species increased in response to external variable) over the first week, while the expression of genes associated with osteogenesis remained unchanged.

Nho *et al.* [60] investigated whether the interaction of MSCs and neural precursor cells (neurospheres) generated from rat cortex with aligned CNTs within a polymer host could selectively enhance stem cell survival. These experiments demonstrated that both MSCs and the neurospheres grew well on the CNT substratum, with both stem cell types directly interacting with the aligned CNTs. The authors suggest that CNTs assisted the proliferation of MSCs and promoted differentiation of neurospheres to neural outcomes.

Laminin-SWNT thin films (fabricated by layer-by-layer assembled composites of SWNTs and laminin) as studied by Kam *et al.* [61] were found to promote the differentiation of neural stem cells (NSCs) and to provide the means for successful excitation of neurons (Figure 4.4). They observed extensive formation of functional neural networks, evident by extensive synaptogenesis *in vitro*. Calcium imaging of neurons differentiated from the NSCs revealed generation of action potentials on the application of a

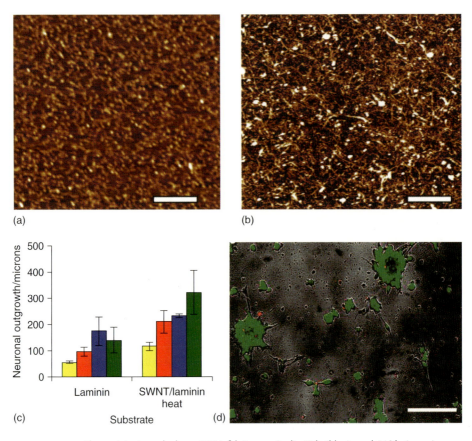

(a)

(b)

(c)

(d)

Figure 4.4 Layer-by-layer (LBL) fabrication. Atomic force microscopy image of (a) one monolayer of laminin on piranha-treated SiO_2 substrate and (b) six bilayers of SWNT/laminin on the same substrate. Scale bars are 1 μm. (c) Distance of outgrowth from neurospheres after 24 h (yellow), 48 h (red), 72 h (blue), and 120 h (green) on laminin-coated slides and heat-treated SWNT/laminin film on slide. (d) Live–dead viability assay on seeded cells, where live cells are stained green and dead cells are red. Scale bar is 200 μm. (Adapted with permission from Kam and coworkers [61].)

lateral current through the SWNT substrate. These results indicated that the laminin-SWNT composite can form the basis of a functional material foundation for neural electrodes.

4.3
Graphene

The fundamental physical properties of graphene electronic devices have been investigated extensively, although relatively little is known about interfacing graphene with biomaterials [62, 63]. To date, there are very few publications detailing the effect of graphene-based material on cellular activity. The ability of graphene to efficiently transfer charge to and from biological molecules may facilitate its use as electrode materials to record cellular events (i.e., neurotransmitter signals) as well as deliver suitable stimulus to surrounding tissue (i.e., application of electrical stimulation to cardiac tissue).

Zhang *et al.* [64] evaluated the *in vitro* toxicity of graphene using PC12 cells and comparing the results to SWNT in similar conditions. A clear difference in the toxic levels between graphene and SWNT was indicated. In addition, they demonstrated that the shape of graphene is directly related to its induced cellular toxicity. Graphene can also be processed into structures, such as papers [65] and films, thus facilitating their use as substrates for sensing or drug delivery. Chen *et al.* [66] prepared electrically conductive graphene paper directly, using the same strategy that has been used to make CNT buckypaper and graphene oxide (GO) papers [65, 67]. They demonstrated that the resulting graphene paper displays a remarkable combination of thermal, mechanical, and electrical properties, while preliminary cytotoxicity tests suggest biocompatibility, making this new material attractive for many potential applications (Figure 4.5).

Cohen-Karni *et al.* [68] constructed a graphene-based field effect transistor (G-FET) to interface with living cells in order to record cellular signals. They reported G-FET conductance signals, recorded from spontaneously beating embryonic chicken cardiomyocytes, which yielded well defined extracellular signals with a signal-to-noise ratio routinely >4. The distinct and complementary capabilities of G-FETs could open up unique opportunities in the field of bioelectronics in the future.

More recently graphene has been used as a substrate to culture stem cells. Stem cell research with graphene has become feasible only with the recent availability of cheap, high-quality, continuous graphene sheets on a large scale. Nayak *et al.* reported that graphene provides a promising biocompatible scaffold that does not hamper the proliferation of human mesenchymal stem cells (hMSCs) and accelerates their specific differentiation into bone cells even in the absence of commonly used additional growth factors such as BMP-2.

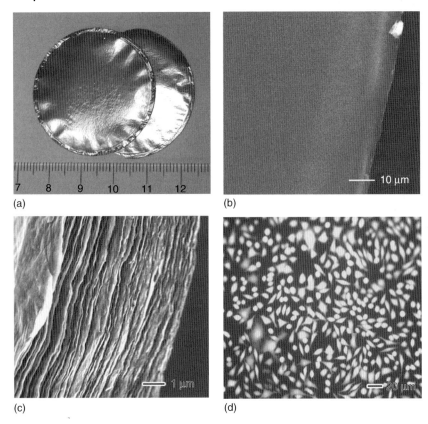

Figure 4.5 (a) Photograph of two pieces of free-standing graphene paper fabricated by vacuum filtration of chemically prepared graphene dispersions, followed by air drying and peeling off the membrane. Front and back surfaces shown. (b) Top view SEM image of a graphene paper sample showing the smooth surface. (c) Side view SEM images of an about 6 µm thick sample at increased magnification. (d) Fluorescence microscopy image of calcein-stained L-929 cells growing on graphene paper. (Adapted with permission from Chen and coworkers [66].)

Induced pluripotent stem cells (iPSC) have been cultured on graphene (G) and graphene oxide (GO). Chen *et al.* showed iPSCs cultured on both G and GO surfaces spontaneously differentiated into ectodermal and mesodermal lineages without significant disparity, but G suppressed the iPSCs differentiation towards the endodermal lineage whereas GO augmented the endodermal differentiation. Human neural stem cells (hNSCs) have been cultured on graphene and demonstrated enhanced neuronal differentiation compared to a glass substrate control. The graphene worked as an excellent cell-adhesion layer during the long-term differentiation process and induced the differentiation of hNSCs more toward neurons than glial cells. In addition Park Chen *et al.* [69] also found that graphene had a good electrical coupling with the differentiated neurons for electrical stimulation.

4.3.1
Carbon-Based Drug Delivery Applications

Many biomedical applications involve the delivery of therapeutic drugs *in vivo*. Research into alternative approaches to drug administration has resulted in utilization of carbon-based materials as drug carriers and/or reservoirs that are capable of delivering their payload at specific sites and in a controlled manner (either via cellular/molecular interaction or via the application of electrical stimulation).

Buckyballs and CNTs have shown promise as delivery vehicles for drugs, including anticancer agents [70]. Paclitaxel-embedded buckysomes (PEBs) are spherical nanostructures in the order of 100–200 nm, composed of the amphiphilic fullerene (AF-1) embedding the anticancer drug paclitaxel inside its hydrophobic pockets. The water-soluble fullerene derivatives enable the uptake of paclitaxel without the need for nonaqueous solvents, which can cause patient discomfort and other unwanted side effects. However, the preliminary studies of Partha *et al.* [70] indicate that PEBs might be capable of delivering even higher amounts of paclitaxel than those delivered via the existing Abraxane® pharmacological strategy. By delivering an increased amount of paclitaxel, it is hoped to reduce infusion times and expect higher tumor uptake, resulting in a greater anticancer efficacy. Another attractive feature of these fullerene-based delivery vectors is that their nanoscale dimensions favor passive targeting, which enables the PEBs to accumulate at tumor sites by entering through leaky vasculature present in the endothelial cell layer of the tumor tissue. In addition, the fullerene moiety can be easily functionalized to attach targeting agents, which facilitate active targeting. PEBs also provide an additional potentially useful feature; that is, the possibility of adding targeting groups to their fullerene moieties.

It has been shown [71] that simple physisorption via π-stacking can be used for loading doxorubicin, a widely used cancer drug, onto nanographene oxide (NGO) functionalized with antibody for selective killing of cancer cells *in vitro*. Owing to its small size, intrinsic optical properties, large specific surface area, low cost, and useful noncovalent interactions with aromatic drug molecules, NGO is a promising new material for biological and medical applications. Recently [72], NGO covalently functionalized with folic acid (FA) has been used to deliver two anticancer drugs simultaneously (doxorubicin – DOX and camptothecin – CPT). It was demonstrated that FA–NGO loaded with the two anticancer drugs showed specific targeting to MCF-7 cells and remarkably high cytotoxicity compared to NGO loaded with either DOX or CPT alone.

Functionalized graphene has provided a means to deliver poorly water-soluble anticancer drugs. Liu *et al.* [73] showed that polyethylene glycolylated-nanographene oxide (NGO-PEG) readily complexes with a water-insoluble aromatic molecule SN38 (a CPT analog) via noncovalent van der Waals interactions. The NGO-PEG-SN38 complex exhibited excellent aqueous solubility and retained the high potency of free SN38 dissolved in organic solvents. The tumor toxicity exceeded that of irinotecan (CPT-11, an FDA-approved SN38 prodrug for colon cancer treatment) by two to three orders of magnitude.

4.4
Conducting Polymers

The ability to rapidly and reversibly switch organic conducting polymers (OCPs) between different oxidation states brings a new dimension to medical bionics. This switching results in dramatic changes in polymer properties such as surface energy, morphology, and even the modulus of the material, imparting physical properties that may be employed to modulate biological behavior(s) in cells. The electrical conductivity and redox activity of the OCPs additionally enable them to function as agents for electrical and/or mechanical stimulation (actuation due to multilayer formats) of target tissues. Furthermore, and of particular importance to their application in medical bionics devices, OCP switching allows uptake and release of dopant ions, which can include bioactive small molecules, protein, and/or nucleic acid species, thus providing an avenue for controlled release of bioactive factors [74].

The development of OCPs for electrodes has thus largely concentrated on their ability to deliver stimulatory paradigms to trigger a regenerative response in target tissues [75–77] or to act as a sensor for electrical activity within the target tissue. It is noteworthy that while some applications may be able to promote regeneration through electrical stimulation, and that these are not, in general, reliant on OCPs, the ability to deliver specific biofactors from OCPs, to modify surface characteristics, and to apply potentials for highly specific control of subsequent regenerative response renders OCPs an important addition to the materials inventory for regenerative electrode-based medical bionics applications.

4.4.1
Neural Applications

Multimodal delivery of synergistic bioactivities that individually target specific aspects of nerve function, protection, or repair have great potential to improve outcomes for diseased or traumatized nerve tissues. The integration of dopant counterions in OCPs imparts an ability to vary their inherent biomimetic character (Figure 4.6), with consequent control of the biological behavior of the cells they come in contact with.

The most immediate application of OCPs to neurobionic devices is in the coating of electrodes used for the electrical stimulation of nerve cells. As such, as for carbon-based electrodes, OCPs also have potential applications within the short term in devices that restore hearing [13], vision [78], and limb movement [37] and control symptoms in Parkinson's disease, dystonia, and chronic pain [14].

The application of OCPs in the area of regenerative bionics for nerve tissues is focused toward an accessory electrode array system that will deliver trophic effects to the peripheral or central nerve environment either through electrical stimulation alone or by electrically stimulated delivery of biological factors that appropriately control the local microenvironment of the nerve tissue [2, 75, 77].

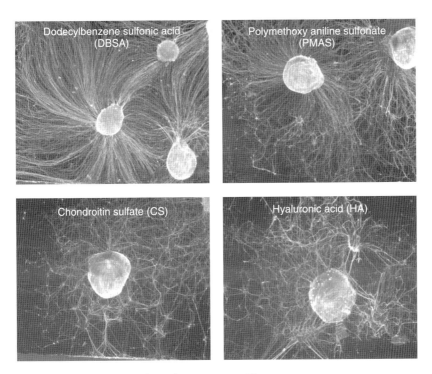

Figure 4.6 Dorsal root ganglia explants grown on PPy doped with biologically derived (HA and CS) and synthetically derived (dodecylbenzenesulfonic acid (DBSA) and PMAS) dopant molecule.

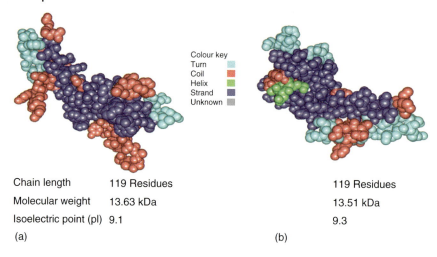

Chain length	119 Residues	119 Residues
Molecular weight	13.63 kDa	13.51 kDa
Isoelectric point (pI)	9.1	9.3
(a)		(b)

Colour key
Turn
Coil
Helix
Strand
Unknown

Figure 4.7 (a) Neurotrophin-3 and (b) brain-derived neurotrophic factor.

In recent work, we have shown that a neurotrophin (NT_3) and brain-derived neurotrophic factor (BDNF) (Figure 4.7) can be incorporated into polypyrrole (PPy) [75] and released using mild electrical stimulation. These materials have been shown to promote a significant increase in neurite extension from spiral ganglia explants [76] (Figure 4.8). We have shown that the growth and differentiation of rat pheochromacytoma (PC12) cells can be assisted by electrically controlled release of a nerve growth factor (NGF) protein [79]. Such capabilities are deemed critical for the survival of auditory neurons in close proximity to an electrode implant. Similar studies with NGF incorporated, but not released, from a

Figure 4.8 Neurite outgrowth is affected by surface chemistry. spiral ganglion neurons (SGN) explants were grown on PPy/pTS (polypyrrole/*para*-toluenesulfonic acid)or PPy/pTS/NT_3, with or without a coating of cell adhesion molecules (CAMs) and in media containing 0 or 40 ng ml^{-1} NT_3. (a) Explants grown on uncoated PPy/pTS exhibited poor neurite outgrowth. (b) The addition of a CAM coating to PPy/pTS-enhanced neurite outgrowth, as did the inclusion of NT_3 in the culture media (c). Compared to PPy/pTS, neurite outgrowth was also improved by incorporating NT_3 into the PPy, exemplified by this explant, grown on PPy/pTS/NT_3 (d). The addition of a CAM coating to PPy/pTS/NT_3 (e) and NT_3 to the media (f) made further dramatic differences to neurite outgrowth. Scale bars are 100 mm. Explant (g) shows significantly increased neural outgrowth when two neurotrophins (NT_3 and BDNF) are delivered *in vitro* under electrical stimulation. Insets in (g) show a schematic of the electrical stimulation setup (left) and an SEM image of the explant tissue sitting on the drug-loaded conducting polymer (right). (Adapted with permission from Thompson and coworkers [76, 84].)

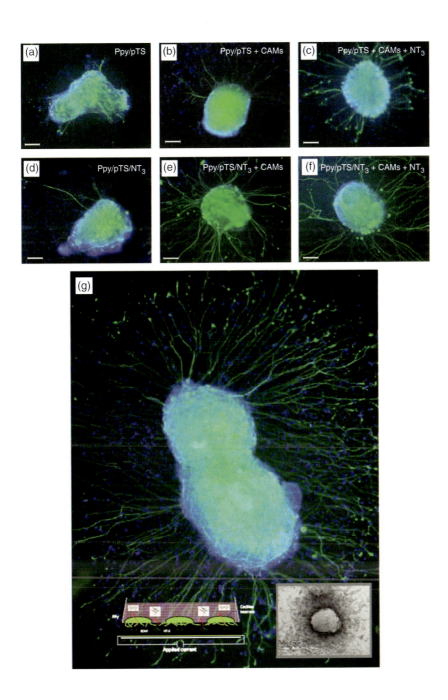

poly(3,4-ethylenedioxythiophene) (PEDOT) matrix have shown to improve the neurite outgrowth of PC12 cells cultured on the substrate, with impedance values of 1.3 kΩ (at 1 Hz) recorded for the composites [80].

However, even without the release of chemical or biochemical proneural molecular species, OCPs have been demonstrated to be potentially very useful materials for tissue engineering scaffolds [81]. Electrical stimulation by OCPs has been shown to promote favorable cell growth, including nerve cells, leading to the development of OCPs for a range of implant applications. Our own studies with primary sensory neuronal explants [77] and those of Langer's group have shown that neurite outgrowth on PPy is facilitated by passage of current through the structure. We have also shown the electrochemical effects on cell growth to be fibronectin dependent [82], a finding recently substantiated by Kotwal and Schmidt [83].

To enhance the nervous tissue–electrode interface, a synthetic protein polymer, SPF, and laminin fragment, CDPGYIGSR, were incorporated through the electrochemical deposition of PPy onto micromachined silicon-based neural recording probes (Figure 4.9a,b) [85]. SPF, with its multiple cell-binding RGD-sequence sites, and laminin fragments were expected to promote cell growth and adhesion on the electrodes. When the neural probes were seeded with rat glial or neuroblastoma cells, the cells preferentially attached to the PPy/SPF and PPy/CDPGYIGSR composite electrode coatings, respectively (Figure 4.9c). The PPy/SPF coatings showed a lower charge capacity compared to control PPy/PSS (polypyrrole/polystyrene sulfonate) coatings (Figure 4.9d) but were still capable of recording good quality voltage signals from single neurons in the cerebellum of guinea pigs. In a different study [86], the same researchers entrapped an additional laminin fragment, RNIAEIIKDI, into PPy, which had a lower impedance and higher charge capacity compared to the CDPGYIGSR fragment mentioned above. Importantly, the PPy/peptide composites showed less astrocyte adhesion compared to bare gold electrodes, which is a promising characteristic for controlling the foreign body response surrounding an electrode.

More recently, PEDOT was grown around living neuronal cells [87]. The cells remained intact and were not lytic/necrotic after the first 24 h following polymerization; however, the detection of activated caspase-3 after 72 h following polymerization indicated an increase in the percentage of apoptotic cells within the PEDOT matrix. The impedance in these composites (1.3 kΩ at 1 Hz), which was lower than bare gold electrodes, was suggested to be due to the electrical activity of the embedded cells that may interfere with the signal transduction between the PEDOT and electrode. By going one step further to integrate conducting polymer with living cells, Richardson-Burns *et al.* [88] introduced a new paradigm for implantable electrodes based on conducting polymers when they polymerized a PEDOT network *in situ* throughout living brain tissue (Figure 4.10). The PEDOT networks were produced by first delivering monomer solution to electrode sites (Teflon coated gold wire) implanted into brain slices. During polymerization, the

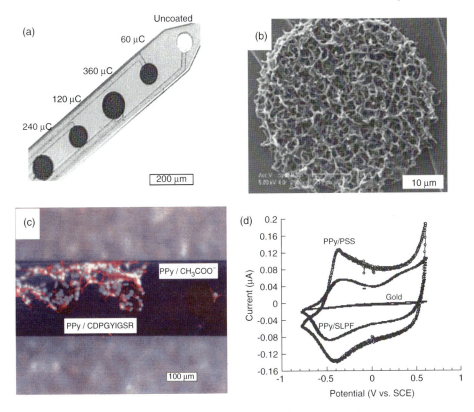

Figure 4.9 (a) Optical micrograph of PPy/PSS-coated five-channel neural probe. (b) SEM images of PPy/SPF-coated electrode sites (total charge passed, 4 μc). (c) Coated neural probe cultured with human neuroblastama cells. (d) Cyclic voltammetry of PPy/SPF-coated electrode in comparison with bare gold and PPy/PSS. (Adapted with permission from Cui and coworkers [85].)

polymer first deposited on the electrode and then grew out into the tissue, forming a diffuse cloud of PEDOT. A network of PEDOT interwoven with cells and their extracellular interstices and components was produced, effectively innervating the nerve tissue. The material had impedances <10 kΩ at 1 Hz when operated as the working electrode. From a medical bionics standpoint, while this is an important aspect of cell/polymer interaction that may find relevance in sensing applications, it must be noted that the integration of the cells within the polymer matrix has the potential to interfere with normal CNS function due to the isolation of those cells from supportive cell types within the polymer. In addition, such integration into CNS tissue does not readily facilitate removal of the polymer mass if or when the embedded cells die, giving rise to the possibility that dead

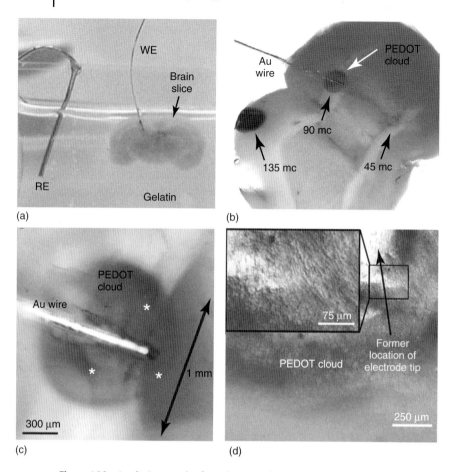

Figure 4.10 (a–d) A network of conducting polymer PEDOT filaments polymerized directly within brain tissue from an implanted electrode. (Adapted with permission from Richardson-Burns and coworkers [87].)

tissue would need to be left behind or that any such removal would damage remaining living tissue.

4.4.2
Muscle Regeneration

The use of different dopants in the formation of OCPs has been shown to generate biomimetic effects in the biological behavior of myoblasts [89]. Furthermore, such varied cellular behaviors associated with dopant variation in excitable cell types (neural cells) *in vitro* have subsequently been observed to translate *in vivo* [90]. Importantly, it is evident that the extent and rate of myoblast to myotube differentiation may be controlled according to the type of OCP that the cells are grown on. For example, films of PPy/CS and PPy/DS (dextran sulfate) and other nonbiological dopants (PPy/poly(2-methoxyaniline-5-sulfonic acid) (PMAS) and PPy/pTS) have been shown to support the differentiation of primary myoblasts into skeletal muscle fibers, whereas PPy/HA did not [89] (Figure 4.11a). Compared to the other dopants, both PPy/chondroitin sulfate (CS) and PPy/HA exhibited higher impedance values in the frequency range for the electrical stimulation of cells (Figure 4.11b), thus providing a further indication of more suitable candidates for use as stimulating electrodes. A concurrent atomic force microscopy (AFM) study on the same composites revealed their nanoscale

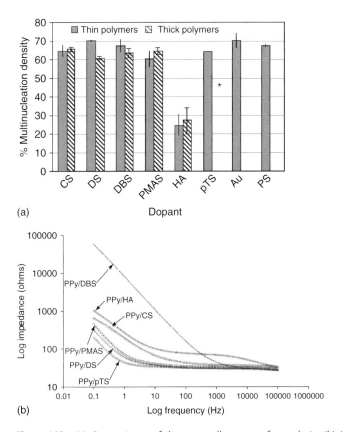

(a)

(b)

Figure 4.11 (a) Comparisons of the percentage of primary myoblast nuclei that were incorporated into multinucleated muscle fibers after 72 h, indicating the degree of skeletal muscle fiber formation on each composite. *Insufficient cells present for analysis. (b) Impedance spectroscopy for thick PPy composites recorded in PBS (pH 7.2) at ±50 mV (vs Ag/AgCl). (Adapted with permission from Gilmore and coworkers [89].)

physical properties tended to group together with films having either a low roughness (≈5 nm root mean square (RMS)), low modulus (≈30 MPa), and a high electromechanically induced strain (≈4.7%) or vice versa (about 30 nm RMS, 1000 MPa, 1.5%) [91]. The passage of current through an OCP substrate has been shown to generate a trophic effect and results in prodifferentiation effects in cultured cells [92]. These effects provide a highly controllable avenue by which to predifferentiate cells on an OCP (PPy or PEDOT) substrate *ex vivo* to an extent that facilitates improved survival and engraftment after implantation of cells in the muscle.

In conjunction with OCPs, biodegradable polymers such as poly(lactide-co-glycolide)-poly(L-lactide) (PLGA:PLA) can be employed to vary microstructural or nanostructural characteristics of OCP surfaces that are able to impose controlled effects on cells before implantation into damaged or diseased muscle [94]. This configuration of substratum showed that wet-spun PLGA:PLA fibers laid on a PPy sheet are an excellent substrate for growth and differentiation of myoblasts into biosynthetic proliferative and partially differentiated myogenic precursor fiber components. Importantly, these myofibers are unbranched and the PLGA:PLA structures can be lifted off the PPy in bundles and surgically implanted into muscle tissue (Figure 4.12). In addition to PLGA:PLA,

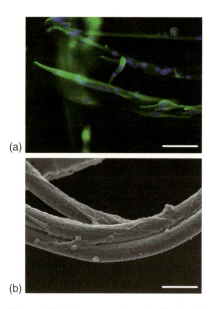

(a)

(b)

Figure 4.12 Linear muscle-seeded biosynthetic fiber constructs. (a) Fluorescence and (b) SEM images of multinucleate myotube-seeded PLGA fibers that have been detached from the polymer substrate. Scale bars are 30 μm. (Adapted with permission from Razal and coworkers [94].)

other biodegradable substrata such as chitosan have shown potential in promoting muscle cell growth [93] and can be readily blended with organic conductors such as PPy [95] to form conducting composite materials suitable for regenerative tissue engineering applications [95, 96].

The two-stage, multidimensional use of OCPs (*ex vivo*) and biodegradable fibers facilitates the possibility that myoregenerative activity in disease situations, such as dystrophic muscle, may be simultaneously stimulated by activation of multiple biochemical routes (e.g., multiple trophic factors). In this scenario, stem cells can be activated to form (i) separate proliferative and (ii) partially differentiated (postmitotic) myoblast biosynthetic muscle tissue components on optimized biodegradable fibers (dimension 1) wound onto OCP substrate (dimension 2) via delivery of a combination of multiple factors and stimulatory paradigms *ex vivo* (stage 1). The biofiber components can then be removed from the OCP substrate and surgically coimplanted as a structure that more closely reflects the cellular niche within functional muscle. Muscle development can be further controlled through the presentation of promyogenic factors incorporated into the biodegradable fibers during their fabrication process (i.e., wet spinning).

Nanostructuring conventional biomaterials profoundly affects regulation of cell adhesion and proliferation in mammalian cells, including muscle [97]. Collectively, these observations give rise to numerous novel applications of OCPs in forming the basis of substrata that could be useful as surfaces that enable control of specific cell growth unprecedented with current technologies. In turn, this opens up exciting possibilities of efficiently growing very specific and potentially homogeneous subpopulations of cells from heterogeneous cell cultures for application in regenerative cell therapies.

4.4.3
Bone

Improved adhesion of osteoblast cells to titanium (Ti) coated with PPy containing a synthetic peptide has been reported [98]. The levels of neonatal rat calvarial osteoblast attachment significantly increased on these peptide-modified PPy substrates compared to unmodified PPy-coated Ti and glass coverslip substrates. Improvements in osteoblast cell attachment, spreading, and proliferation have been observed for similar PPy/Ti substrates [99]. New surfaces based on Ti coated on poly(pyrrole-3-acetic acid) coatings provide a further platform to create biofunctionalized interfaces [100]. Reactive carboxylic surface groups of the polymer offer covalent tailoring with peptides, with the films showing good biocompatibility for mouse bone marrow cells. The incorporation of polysaccharides, namely, heparin, chondrotin-4-sulfate, and hyaluronic acid (HA), are another means of assuring suitable chemistry, as demonstrated by positive growth of MC3T3-E1 osteoblasts [101]. The ability of rat MSCs to differentiate toward an osteoblastic phenotype can also be modulated through culturing on PPy films chemically polymerized in polystyrene dishes [102]. By varying the monomer concentration, a wide spectrum of cell attachment characteristics from excellent cell attachment to the complete inability to adhere can be achieved. In particular, specific concentrations demonstrated superior induction of mesenchymal stem cell osteogenicity. Studies have recently progressed to electrically stimulated release of antibiotics (penicillin/streptomycin) and anti-inflammatory drugs (dexamethasone) that hold great potential to support osteoblast proliferation and fight bacterial infection [103], the latter of which is critical for implants addressing bone applications. Direct electrical stimulation of osteoblasts [104], in addition to controlled release of drugs [105], is proving promising for cell growth in much the same way as the above-mentioned neural applications.

4.4.4
Stem Cells

The ability to integrate a conducting polymer coating, capable of releasing chemical factors [105], onto Ti materials opens up the possibility of recruiting and differentiating MSCs to assist in the bone rebuilding process, as described above. The adhesion of stem cells to the substrate is another critical parameter as this regulates cell density, which further presides over their differentiation toward mature cells. Elegant control of NSC adhesion by fabricating an "electronic polymer surface switch" that modulates the oxidation state of thin PEDOT:Tosylate films using electrical stimulation has been reported [106]. Even without electrical control, other types of progenitor/stem cells including human embryonic stem cells [107] and mouse muscle myoblasts [89] have shown enhanced differentiation in response to the type of dopant incorporated in the polymer substrate. The theme of controlling stem cell fate is expected to grow rapidly, as OCPs appear to have all the necessary attributes, such as controllable topography and stiffness, and the ability to deliver chemical and electrical cues.

4.5
Toxicity

The issue of toxicity of organic conductors will need to be addressed on a number of fronts, and sooner rather than later, as a growing body of multidisciplinary research teams rapidly develops these materials ever closer to clinical applications. As with the current FDA-approved materials used in neuroprosthetics, an initial battery of information will be essential for validating the use of organic conductors in downstream human studies. Basic characterization studies on the gamut of materials properties, biocompatibility, cellular–materials interactions, and long-term stability, as described throughout this and previous chapters, facilitate the early stages of material screening and identification of suitable materials configurations (e.g., nerve conduits) to ensure efficacy, justify financial costs, and gain ethics approval as studies progress toward the use of animals. Armed with this information and that obtained from *in vivo* animal studies, the issue of additional cytotoxicity data becomes more critical as developed materials confront risk–benefit analysis, FDA approval processes, and human ethics committees before they are ultimately deemed safe for human testing.

Additional resistance to modifying current devices by end users and industry may be met because of the costs and lengthy times involved in approving the use of new materials such as is the case for organic conductors that are envisaged as replacements for already approved and established electrode components. Undertaking specifically designed toxicity studies that run in parallel, which is often not the case, is therefore well-advised to fully complement a speedy translation of organic-conductor-based electrodes from the benchtop to industry. The perception of the material's toxicity in the scientific community and its foray into the public arena also bears heavily on the future outlook of the material. This is best exemplified by the current status on the toxicity of CNTs, and nanoparticles in general [108–110]. Recently, an agreement has been reached toward the potential toxicity of CNTs, but there are still many conflicting reports believed to be associated with the different effects elicited from the vast array of CNT configurations ranging from pure, functionalized, dispersed, aggregated, or immobilized CNTs and the way in which they are used [111]. It is relevant at this point to emphasize that the final configuration of the organic conductor will strongly determine its toxicity – not necessarily the starting material. For example, CNTs immobilized on a substrate (e.g., CNT electrode arrays) would appear to be of far less concern than freely dispersed CNTs that are commonly linked to cytotoxicity. Similar considerations should apply to graphene and OCPs, particularly in light of medical bionics devices mainly requiring that these materials be configured as a fixed coating, electrode or 3D structure for electrical stimulation or recording purposes. In the case of conducting polymers, their fabrication is predominately in the form of adherent coatings or films, which to date have proved to be noncytotoxic *in vivo* [112, 113]. Future toxicity studies in this area may also be better served by focusing on the effect of products (e.g., chemical moieties, dopants, and monomers) released from immobilized organic conducting structures in the advent of degradation [114].

4.6
Sterilization

Just as with all other materials to be implanted into living systems, organic conductors to be used for medical bionics applications need to be able to withstand conditions imposed on them by sterilization procedures. Sterilization destroys any microbiota, such as bacteria or fungi in both adult and/or sporular life phases, that may be attached to a given surface and may be pathogenic if implanted with the device. Some sterilization protocols have an additional advantage in being able to destroy viral contaminants, again rendering the sterilized surface free of viral pathogens. In general, sterilization involves application of physical factors such as heat, high pressure, and filtration or molecular factors such as chemicals or irradiation, either on their own or in combinations, to generate an environment that renders any life-forms resident on the materials' surface unable to persist. As such, sterilization methods have an inordinate capacity to interfere with the integrity of any given material that coats the surface of a medical bionic implant, and the specific method to be used for any given surface will necessarily need to be a trade-off between effectiveness and the propensity to change the nature of the implant surface.

In addition to the effects of any given sterilization method on the material and potential pathogens, the chosen sterilization method needs to involve consideration of its effect on any proteinaceous residues on the material surface and whether any such contaminant will be left in a state that benefits surface pathogen survival of the sterilization process. For example, any microparticulate substance left on the material surface may be denatured (e.g., proteinaceous) to a form that facilitates adhesion to the surface and subsequent provision of shelter from the sterilization effects or nourishment of any surviving pathogens poststerilization. In more recent times, greater focus of sterilization techniques has been on the control of (relatively low frequency) transmissible prions that cause spongiform encephalopathies such as Cruetzfeld–Jakob disease [115].

4.6.1
Physical Methods of Sterilization

- **Filtration:** Filtration can be a useful form of sterilization, although it is of limited value in the eradication of viral or prion contamination. Filtration involves the passage of materials in liquid form through filters with pore sizes that retain particles within the size of bacterial and fungal organisms. This method is of great advantage for the sterilization of OCP components (dopants and monomer/repeating unit) solutions before their polymerization in a clean-room environment. Using filtration, there is little possibility of changing the physicochemical properties of the

OCP after sterilization, thereby providing an excellent means to promote reproducibility of sterilized products. However, filtration will not remove viral or prion contamination.

- **Dry heat:** Dry heat sterilization involves exposure to hot air in the absence of water vapor [116, 117]. In this process, heat is absorbed layer by layer into the body of the material being sterilized until the point where the entire volume of the material reaches a uniform temperature sufficient to kill all living organisms that may be present in or on the surface of the material. Because heat energy does not transfer across any given surface interface as effectively in absence of water, dry heat sterilization temperatures need to be higher than those required in the presence of water. Thus, materials sterilized by dry heat are heated to temperatures of $160–170\,^{\circ}C$, at which they are maintained for 2 h or down to 1 h respectively. At these temperatures, the protein components of micro-organisms are denatured, thereby neutralizing the organism. Materials that have been sterilized by dry heat are protected from degradation or damage arising from the effects of wetting. However, the relatively high temperatures and long exposure times required for effective dry heat sterilization have a propensity to impose significant change in the surface properties and therefore the biomimetic properties of OCPs.

- **Wet heat (steam):** In contrast to dry sterilization techniques, materials that are sterilized using heated water require exposure to lower temperatures for significantly lower time frames in order to be rendered safe for implantation as part of medical devices. However, the "usual" exposure of materials to water (i.e., steam) heated to $121\,^{\circ}C$ for $15–20\,min$ or up to $134\,^{\circ}C$ for between 3 and 5 min means that any component likely to be damage by exposure to water cannot be sterilized by this method. This complication is exacerbated if more heat-resistant organisms such as bacterial spores and prions are considered as targets for the sterilization process. In these cases, the exposure of materials to temperatures in the higher end of the effective sterilization scale for much longer times (e.g., $132\,^{\circ}C/60\,min$ or $134\,^{\circ}C/20\,min$) renders them much more susceptible to heat damage than conventional steam sterilization applications. It is also a fact that because of the high working temperatures and pressures generated by this process, autoclaves tend to undergo more rapid deterioration (e.g., rusting), a process augmented if they are used for processing liquids containing salts, which in turn also leads to a somewhat variable chemical environment if the water used in the autoclave has salts dissolved in it. For these reasons, steam sterilization is capable of producing highly variable effects on polymeric materials.

4.6.2
Irradiation

Sterilization can be achieved by irradiating materials with X-rays, γ-rays, electron beams, or subatomic particles [118].

- **X-Ray irradiation:** The high ionizing energy (up to 5–7 MeV) of X-rays (bremsstrahlung) facilitates irradiation and thereby sterilization of large volumes of high-density materials at a time. X-ray sterilization is not dependent on chemicals or radioactive substrate but rather uses electrical stimulation of tantalum.
- **γ Irradiation:** γ-rays are a form of short wavelength electromagnetic radiation generated through radioactive decay of isotopes such as ^{60}Co, which sterilize materials via damage to the DNA (and other essential biomolecules) of contaminating organisms. The deep penetration of materials by ionizing γ-rays arises through high frequency ($>10^{19}$ Hz) and has great potential to alter the chemical structure of OCP materials. Largely due to the high energies (>100 keV) generated by γ-rays, the molecular integrity of irradiated polymers is compromised by γ-rays. Alternatively, γ-irradiation does not involve heat, and so is able to be used on materials that are labile under conditions of high heat.

4.6.3
Electron Beam (E-Beam)

Electron beams (E-beams, between 0.75 and 10 MeV), formed by electrically stimulated tungsten filaments under vacuum in electron accelerators, are currently used mostly for sterilization of polymer materials used for construction of or packaging medical products. E-beam sterilization delivers a well-controlled narrowband dosage that kills contaminating biota in a manner similar to γ-irradiation, by modifying biomolecules (such as DNA) that are essential to life. E-beam irradiation is also used for the modification of certain polymers, introducing cross-linkages and, in other cases, breaking down linkages in polymers in a manner that is determined by the polymer structure. In general, E-beam sterilization is less destructive than γ-irradiation, but this is dependent on the type of material being irradiated. This is largely due to the fact that E-beams generate a higher dosage rate than does γ-radiation (or even X-ray), thus requiring a lower time of exposure to promote effective sterilization. However, the capability of E-beams to initiate chemical reactions by providing appropriate activation energies means that they are also capable of inducing unwanted chemical changes in polymeric components.

4.6.4
Ultraviolet (UV) Light Irradiation

Ultraviolet (UV) light is absorbed by the double bonds of many compounds such as DNA and proteins. As such, it is useful for the neutralization of microorganisms but because of its relatively low penetrating capability is only useful for the treatment of material surfaces. The reactivity of UV radiation with polymer backbones makes it unsuitable for use with some polymers.

4.6.4.1 Plasma Sterilization

Plasma sterilization utilizing UV photons and chemically reactive radicals ($O^{\bullet}$ and $OH^{\bullet}$) promotes sterilization at temperatures below $50\,^{\circ}C$ and, as such, is potentially even more reactive with the carbon backbones than just UV light [119]. However, the additional use of radicals (oxidizing agents) facilitates highly effective erosion properties with regard to contaminants such as spores, which are otherwise impervious to UV photons alone.

4.6.5
Chemical Methods of Sterilization

Dry chemical sterilization (DCS) exposing material to hydrogen peroxide (H_2O_2: 30–35% v:v) for as little as 6–10 s under low pressure and temperature (10–15 $^{\circ}C$) is an effective method for the reduction of bacterial contamination, but in some sterilization applications, H_2O_2 concentrations as high as 90% v:v are commonly used for as long as 30 min. DCS works via vaporization and subsequent thermal disassociation of H_2O_2 (through a mildly exothermic process), thereby generating oxygen (mainly) free radicals that kill resident contaminating biota more or less immediately on contact. DCS promotes between six and eight orders of magnitude reduction in bacterial-colony-forming units after treatment and is effective in the treatment of contamination with bacteria (including spores) virus, fungi, and prions. While conventional medical equipment is well served by DCS, the chemical nature of the process is potentially a disadvantage for the treatment of medical bionic implant coatings. Nevertheless, DCS is useful for materials that are sensitive to higher temperatures but impervious to chemical attack from H_2O_2.

Hydrogen peroxide can also be mixed with formic acid as needed in the Endoclens device for sterilization of endoscopes. This device has two independent asynchronous bays and cleans (in warm detergent with pulsed air), sterilizes, and dries endoscopes automatically in 30 min. Studies with synthetic soil with bacterial spores showed the effectiveness of this device.

4.6.6
Ethylene Oxide (EtO)

Sterilization with highly flammable/explosive ethylene oxide (EtO) gas involves exposure of materials to between 200 and 800 mg l^{-1} of EtO gas at a humidity of at least 30%, between 30 and 60 °C for at least 3 h. EtO is thus useful for materials sensitive to higher temperatures (>60 °C) or irradiation with ionizing radiation. EtO is effective against viruses, viable and spore forms, and fungi. Processes using EtO are highly toxic, involving contamination of air with residual EtO and associated CFC contaminants; EtO is explosive from concentrations as low as 3% and toxic at concentrations as low as 500–800 ppm. Nevertheless, overall, EtO is ideal for the sterilization of many materials and, because of its broad spectrum of effect and noninvolvement of heat, is by far the most common method used for sterilization in general.

4.6.7
Ozone (O₃)

Ozone is a powerful oxidant that is effective against viral, bacterial, and prion pathogens. Ozone gas is highly reactive, with a toxicity concentration as low as 5 ppm, and promotes effective sterilization after 4.5 h of exposure time. It is used in industrial settings to sterilize water and air, and as a disinfectant for surfaces. It has the benefit of being able to oxidize most organic matter. On the other hand, it is a toxic and unstable gas that must be produced on-site, so it is not practical for use in many settings.

4.6.8
Bleach (Sodium Hypochlorite)

Chlorine in solution (sodium hypochlorite, NaOCl; ~5%) is a strong oxidizing agent and an effective sterilizing chemical for bacterial (including many spores), viral, fungal, and prion contaminants. NaOCl kills organisms largely on contact, but ideally it is best left to react for at least 20 min. As a strong oxidizing agent, NaOCl is also highly corrosive and reacts with metals, acids, and spontaneously decomposes when exposed to air and/or heated [120].

4.6.9
Glutaraldehyde and Formaldehyde

In solution, glutaraldehyde and formaldehyde can sterilize immersed materials against bacteria (some spores) and viruses. Aldehydes are relatively ineffective against prions. The necessity to immerse materials to be sterilized for long periods of time renders this a messy methodology that has similar disadvantages to wet (steam) and wet chemical-based techniques. The utility of glutaraldehyde and formaldehyde may be augmented by low-temperature steam, which greatly reduces the time required for treatment and facilitates the penetration of aldehydes as sterilization agents.

4.6.10
Ortho-Phthalaldehyde (OPA)

Ortho-phthalaldehyde (OPA) is used in a 0.55% solution, and is more effective against mycobactericidal activity than formaldehyde or glutaraldehyde, with increased efficacy against some spores not treated effectively with other aldehydes.

While the list presented here is not exhaustive, it represents the majority of methodologies employed in the sterilization of materials to be used for human medical/surgical purposes. From this context, the advent of conducting carbon materials' application to present and future implantable medical bionics devices dictates concomitant evolution in fabrication protocols that promote maximal structural and functional integrity within the material while limiting their capacity to introduce unwanted biota into the implanted tissue.

References

1. Hargrove, L.J., Huang, H., Schultz, A.E., Lock, B.A., Lipschutz, R., and Kuiken, T.A. (2009) Toward the development of a neural interface for lower limb prosthesis control. *Conf. Proc. IEEE Eng. Med. Biol. Soc.*, **2009**, 2111–2114.

2. Richardson, R.T., Wise, A.K., Thompson, B.C., Flynn, B.O., Atkinson, P.J., Fretwell, N.J., Fallon, J.B., Wallace, G.G., Shepherd, R.K., Clark, G.M. *et al.* (2009) Polypyrrole-coated electrodes for the delivery of charge and neurotrophins to cochlear neurons. *Biomaterials*, **30**, 2614–2624.

3. Breukers, R.D., Gilmore, K.J., Kita, M., Wagner, K.K., Higgins, M.J., Moulton, S.E., Clark, G.M., Officer, D.L., Kapsa, R.M., and Wallace, G.G. (2010) Creating conductive structures for cell growth: growth and alignment of myogenic cell types on polythiophenes. *J. Biomed. Mater. Res. A*, **95**, 256–268.

4. Blazewicz, M. (2001) Carbon materials in the treatment of soft and hard tissue injuries. *Eur. Cell Mater.*, **2**, 21–29.

5. Williams, D.F. and Roaf, R. (1973) *Implants in Surgery*, W. B. Saunders Co, London.

6. Wintermantel, E., Mayer, J., and Goehring, T.N. (2006) in *Concise Encyclopedia of Composite Materials* (ed. A. Mortensen), Elsevier, Amsterdam, London, p. 1050.

7. Masini, M., Lazzari, M., Lorenzoni, R., Domicelli, A.M., Micheletti, A., Dianda, R., and Masini, G. (1996) Activated pyrolytic carbon tip pacing leads: an alternative to steroid-eluting pacing leads? *Pacing Clin. Electrophysiol.*, **19**, 1832–1835.

8. Ripart, A. and Mugica, J. (1983) Electrode-heart interface: definition of the ideal electrode. *Pacing Clin. Electrophysiol.*, **6**, 410–421.

9. Mugica, J., Henry, L., Attuel, P., Lazarus, B., and Duconge, R. (1986) Clinical experience with 910 carbon tip leads: comparison with polished platinum leads. *Pacing Clin. Electrophysiol.*, **9**, 1230–1238.

10. Garberoglio, B., Inguaggiato, B., Chinaglia, B., and Cerise, O. (1983) Initial results with an activated pyrolytic carbon tip electrode. *Pacing Clin. Electrophysiol.*, **6**, 440–448.

11. Pioger, G., Carberoglio, B., and Antonioli, G.E. (1991) in *Proceedings of the 2nd European Conference on Pacemaker Leads* (ed. A.E. Aubert), Elsevier Science Ltd, Amsterdam, pp. 475–480.

12. Alt, E.U., Fotuhi, P.C., Callihan, R.L., Rollins, D.L., Mestre, E., Combs, M.P., Smith, W.M., and Ideker, R.E. (1995) Improved defibrillation threshold with a new epicardial carbon electrode compared with a standard epicardial titanium patch. *Circulation*, **91**, 445–450.

13. Clark, G.M. (2006) The multiple-channel cochlear implant: the interface between sound and the central nervous system for hearing, speech, and language in deaf people – a personal perspective. *Philos. Trans. R. Soc. B: Biol. Sci.*, **361**, 791–810.

14. Benabid, A.L. (2003) Deep brain stimulation for Parkinson's disease. *Curr. Opin. Neurobiol.*, **13**, 696–706.

15. Greenberg, B.D., Askland, K.D., and Carpenter, L.L. (2008) The evolution of deep brain stimulation for neuropsychiatric disorders. *Front. Biosci.*, **13**, 4638–4648.

16. Su, H.C., Lin, C.M., Yen, S.J., Chen, Y.C., Chen, C.H., Yeh, S.R., Fang, W., Chen, H., Yao, D.J., and Chang, Y.C. *et al.* (2010) A cone-shaped 3D carbon nanotube probe for neural recording. *Biosens. Bioelectron.*, **26**, 220–227.

17. Lin, C.M., Lee, Y.T., Yeh, S.R., and Fang, W. (2009) Flexible carbon nanotubes electrode for neural recording. *Biosens. Bioelectron.*, **24**, 2791–2797.

18. Harrison, B.S. and Atala, A. (2007) Carbon nanotube applications for tissue engineering. *Biomaterials*, **28**, 344–353.

19. Armstrong-James, M., Fox, K., Kruk, Z.L., and Millar, J. (1981) Quantitative ionophoresis of catecholamines using multibarrel carbon fibre microelectrodes. *J. Neurosci. Methods*, **4**, 385–406.

20. Kuras, A. and Gutmaniene, N. (2000) Technique for producing a carbon-fibre microelectrode with the fine recording tip. *J. Neurosci. Methods*, **96**, 143–146.

21. Anderson, C.W. and Cushman, M.R. (1981) A simple and rapid method for making carbon fiber microelectrodes. *J. Neurosci. Methods*, **4**, 435–436.

22. Millar, J. and Pelling, C.W.A. (2001) Improved methods for construction of carbon fibre electrodes for extracellular spike recording. *J. Neurosci. Methods*, **110**, 1–8.

23. de Asis, E.D., Nguyen-Vu, T.D., Arumugam, P.U., Chen, H., Cassell, A.M., Andrews, R.J., Yang, C.Y., and Li, J. Jr. (2009) High efficient electrical stimulation of hippocampal slices with vertically aligned carbon nanofiber microbrush array. *Biomed. Microdevices*, **11**, 801–808.

24. Inagaki, K., Heiney, S.A., and Blazquez, P.M. (2009) Method for the construction and use of carbon fiber multibarrel electrodes for deep brain recordings in the alert animal. *J. Neurosci. Methods*, **178**, 255–262.

25. Zimmerman, M., Parsons, J.R., Alexander, H., and Weiss, A.B. (1984) The electrical stimulation of bone using a filamentous carbon cathode. *J. Biomed. Mater. Res.*, **18**, 927–938.

26. Krishnan, A., Dujardin, E., Ebbesen, T.W., Yianilos, P.N.,

and Treacy, M.M.J. (1998) Young's modulus of single-walled nanotubes. *Phys. Rev. B*, **58**, 14013.

27. Falvo, M.R., Clary, G.J., Taylor, R.M. II, Chi, V., Brooks, F.P. Jr., Washburn, S., and Superfine, R. (1997). Bending and buckling of carbon nanotubes under large strain. *Nature*, **389**, 582–584.

28. Saito, N., Usui, Y., Aoki, K., Narita, N., Shimizu, M., Hara, K., Ogiwara, N., Nakamura, K., Ishigaki, N., and Kato, H. *et al.* (2009) Carbon nanotubes: biomaterial applications. *Chem. Soc. Rev.*, **38**, 1897–1903.

29. Flahaut, E., Durrieu, M.C., Remy-Zolghadri, M., Bareille, R., and Baquey, C. (2006) Investigation of the cytotoxicity of CCVD carbon nanotubes towards human umbilical vein endothelial cells. *Carbon*, **44**, 1093–1099.

30. Jan, E. and Kotov, N.A. (2007) Successful differentiation of mouse neural stem cells on layer-by-layer assembled single-walled carbon nanotube composite. *Nano Lett.*, **7**, 1123–1128.

31. Chao, T.I., Xiang, S., Chen, C.S., Chin, W.C., Nelson, A.J., Wang, C., and Lu, J. (2009) Carbon nanotubes promote neuron differentiation from human embryonic stem cells. *Biochem. Biophys. Res. Commun.*, **384**, 426–430.

32. Mooney, E., Dockery, P., Greiser, U., Murphy, M., and Barron, V. (2008) Carbon nanotubes and mesenchymal stem cells: biocompatibility, proliferation and differentiation. *Nano Lett.*, **8**, 2137–2143.

33. Mattson, M.P., Haddon, R.C., and Rao, A.M. (2000) Molecular functionalization of carbon nanotubes and use as substrates for neuronal growth. *J. Mol. Neurosci.*, **14**, 175–182.

34. Mazzatenta, A., Giugliano, M., Campidelli, S., Gambazzi, L., Businaro, L., Markram, H., Prato, M., and Ballerini, L. (2007) Interfacing neurons with carbon nanotubes: electrical signal transfer and synaptic stimulation in cultured brain circuits. *J. Neurosci.*, **27**, 6931–6936.

35. Wilson, B.S., Lawson, D.T., Muller, J.M., Tyler, R.S., and Kiefer, J. (2003) Cochlear implants: some likely next steps. *Annu. Rev. Biomed. Eng.*, **5**, 207–249.

36. Weiland, J.D., Liu, W., and Humayun, M.S. (2005) Retinal prosthesis. *Annu. Rev. Biomed. Eng.*, **7**, 361–401.

37. Navarro, X., Krueger, T.B., Lago, N., Micera, S., Stieglitz, T., and Dario, P. (2005) A critical review of interfaces with the peripheral nervous system for the control of neuroprostheses and hybrid bionic systems. *J. Peripher. Nerv. Syst.*, **10**, 229–258.

38. Merrill, D.R., Bikson, M., and Jefferys, J.G. (2005) Electrical stimulation of excitable tissue: design of efficacious and safe protocols. *J. Neurosci. Methods*, **141**, 171–198.

39. Brummer, S.B. and Turner, M.J. (1977) Electrochemical considerations for safe electrical stimulation of the nervous system with platinum electrodes. *IEEE Trans. Biomed. Eng.*, **24**, 59–63.

40. Loeb, G.E., Peck, R.A., and Martyniuk, J. (1995) Toward the ultimate metal microelectrode. *J. Neurosci. Methods*, **63**, 175–183.

41. Hu, H., Ni, Y., Mandal, S.K., Montana, V., Zhao, B., Haddon, R.C., and Parpura, V. (2005) Polyethyleneimine functionalized single-walled carbon nanotubes as a substrate for neuronal growth. *J. Phys. Chem. B*, **109**, 4285–4289.

42. Lovat, V., Pantarotto, D., Lagostena, L., Cacciari, B., Grandolfo, M., Righi, M., Spalluto, G., Prato, M., and Ballerini, L. (2005) Carbon nanotube substrates boost neuronal electrical signaling. *Nano Lett.*, **5**, 1107–1110.

43. Bekyarova, E., Ni, Y., Malarkey, E.B., Montana, V., McWilliams, J.L., Haddon, R.C., and Parpura, V. (2005) Applications of carbon

nanotubes in biotechnology and biomedicine. *J. Biomed. Nanotechnol.*, **1**, 3–17.

44. Liopo, A.V., Stewart, M.P., Hudson, J., Tour, J.M., and Pappas, T.C. (2006) Biocompatibility of native and functionalized single-walled carbon nanotubes for neuronal interface. *J. Nanosci. Nanotechnol.*, **6**, 1365–1374.

45. McKenzie, J.L., Waid, M.C., Shi, R., and Webster, T.J. (2004) Decreased functions of astrocytes on carbon nanofiber materials. *Biomaterials*, **25**, 1309–1317.

46. Silva, G.A. (2006) Nanomedicine: seeing the benefits of ceria. *Nat. Nanotechnol.*, **1**, 92–94.

47. Correa-Duarte, M.A., Wagner, N., Rojas-Chapana, J., Morsczeck, C., Thie, M., and Giersig, M. (2004) Fabrication and biocompatibility of carbon nanotube-based 3D networks as scaffolds for cell seeding and growth. *Nano Lett.*, **4**, 2233–2236.

48. Gabay, T., Jakobs, E., Ben-Jacob, E., and Hanein, Y. (2005) Engineered self-organization of neural networks using carbon nanotube clusters. *Physica A: Stat. Mech. Appl.*, **350**, 611–621.

49. Ayutsede, J., Gandhi, M., Sukigara, S., Ye, H., Hsu, C.M., Gogotsi, Y., and Ko, F. (2006) Carbon nanotube reinforced Bombyx mori silk nanofibers by the electrospinning process. *Biomacromolecules*, **7**, 208–214.

50. Keefer, E.W., Botterman, B.R., Romero, M.I., Rossi, A.F., and Gross, G.W. (2008) Carbon nanotube coating improves neuronal recordings. *Nat. Nanotechnol.*, **3**, 434–439.

51. MacDonald, R.A., Laurenzi, B.F., Viswanathan, G., Ajayan, P.M., and Stegemann, J.P. (2005) Collagen-carbon nanotube composite materials as scaffolds in tissue engineering. *J. Biomed. Mater. Res. A*, **74**, 489–496.

52. Garibaldi, S., Brunelli, C., Bavastrello, V., Ghigliotti, G., and Nicolini, C. (2006) Carbon nanotube biocompatibility with cardiac muscle cells. *Nanotechnology*, **17**, 391.

53. Saito, N., Usui, Y., Aoki, K., Narita, N., Shimizu, M., Ogiwara, N., Nakamura, K., Ishigaki, N., Kato, H., Taruta, S. *et al.* (2008) Carbon nanotubes for biomaterials in contact with bone. *Curr. Med. Chem.*, **15**, 523–527.

54. Yadav, S.K., Bera, T., Saxena, P.S., Maurya, A.K., Garbyal, R.S., Vajtai, R., Ramachandrarao, P., and Srivastava, A. (2010) MWCNTs as reinforcing agent to the hap-gel nanocomposite for artificial bone grafting. *J. Biomed. Mater. Res. A*, **93**, 886–896.

55. Zhao, B., Hu, H., Mandal, S.K., and Haddon, R.C. (2005) A bone mimic based on the self-assembly of hydroxyapatite on chemically functionalized single-walled carbon nanotubes. *Chem. Mater.*, **17**, 3235–3241.

56. Supronowicz, P.R., Ajayan, P.M., Ullmann, K.R., Arulanandam, B.P., Metzger, D.W., and Bizios, R. (2002) Novel current-conducting composite substrates for exposing osteoblasts to alternating current stimulation. *J. Biomed. Mater. Res.*, **59**, 499–506.

57. Zanello, L.P., Zhao, B., Hu, H., and Haddon, R.C. (2006) Bone cell proliferation on carbon nanotubes. *Nano Lett.*, **6**, 562–567.

58. McCullen, S.D., Stevens, D.R., Roberts, W.A., Clarke, L.I., Bernacki, S.H., Gorga, R.E., and Loboa, E.G. (2007) Characterization of electrospun nanocomposite scaffolds and biocompatibility with adipose-derived human mesenchymal stem cells. *Int. J. Nanomedicine*, **2**, 253–263.

59. Tay, C.Y., Gu, H., Leong, W.S., Yu, H., Li, H.Q., Heng, B.C., Tantang, H., Loo, S.C.J., Li, L.J., and Tan, L.P. (2010) Cellular behavior of human mesenchymal stem cells

cultured on single-walled carbon nanotube film. *Carbon*, **48**, 1095–1104.

60. Nho, Y., Kim, J.Y., Khang, D., Webster, T.J., and Lee, J.E. (2010) Adsorption of mesenchymal stem cells and cortical neural stem cells on carbon nanotube/polycarbonate urethane. *Nanomedicine (Lond)*, **5**, 409–417.

61. Kam, N.W., Jan, E., and Kotov, N.A. (2009) Electrical stimulation of neural stem cells mediated by humanized carbon nanotube composite made with extracellular matrix protein. *Nano Lett.*, **9**, 273–278.

62. Ohno, Y., Maehashi, K., Yamashiro, Y., and Matsumoto, K. (2009) Electrolyte-gated graphene field-effect transistors for detecting pH and protein adsorption. *Nano Lett.*, **9**, 3318–3322.

63. Mohanty, N. and Berry, V. (2008) Graphene-based single-bacterium resolution biodevice and DNA transistor: interfacing graphene derivatives with nanoscale and microscale biocomponents. *Nano Lett.*, **8**, 4469–4476.

64. Zhang, Y., Ali, S.F., Dervishi, E., Xu, Y., Li, Z., Casciano, D., and Biris, A.S. (2010) Cytotoxicity effects of graphene and single-wall carbon nanotubes in neural phaeochromocytoma-derived PC12 cells. *ACS Nano*, **4**, 3181–3186.

65. Dikin, D.A., Stankovich, S., Zimney, E.J., Piner, R.D., Dommett, G.H., Evmenenko, G., Nguyen, S.T., and Ruoff, R.S. (2007) Preparation and characterization of graphene oxide paper. *Nature*, **448**, 457–460.

66. Chen, H., Müller, M.B., Gilmore, K.J., Wallace, G.G., and Li, D. (2008) Mechanically strong, electrically conductive, and biocompatible graphene paper. *Adv. Mater.*, **20**, 3557–3561.

67. Park, S., Lee, K.S., Bozoklu, G., Cai, W., Nguyen, S.T., and Ruoff, R.S. (2008) Graphene oxide papers modified by divalent ions-enhancing mechanical properties via chemical cross-linking. *ACS Nano*, **2**, 572–578.

68. Cohen-Karni, T., Qing, Q., Li, Q., Fang, Y., and Lieber, C.M. (2010) Graphene and nanowire transistors for cellular interfaces and electrical recording. *Nano Lett.*, **10**, 1098–1102.

69. Park, S.Y., Park, J., Sim, S.H., Sung, M.G., Kim, K.S., Hong, B.H., and Hong, S. (2011) and *Adv. Mater.*, **23**, H263–H267.

70. Partha, R., Mitchell, L.R., Lyon, J.L., Joshi, P.P., and Conyers, J.L. (2008) Buckysomes: fullerene-based nanocarriers for hydrophobic molecule delivery. *ACS Nano*, **2**, 1950–1958.

71. Sun, X., Liu, Z., Welsher, K., Robinson, J.T., Goodwin, A., Zaric, S., and Dai, H. (2008) Nano-Graphene oxide for cellular imaging and drug delivery. *Nano Res.*, **1**, 203–212.

72. Zhang, L., Xia, J., Zhao, Q., Liu, L., and Zhang, Z. (2010) Functional graphene oxide as a nanocarrier for controlled loading and targeted delivery of mixed anticancer drugs. *Small*, **6**, 537–544.

73. Liu, Z., Robinson, J.T., Sun, X., and Dai, H. (2008) PEGylated nanographene oxide for delivery of water-insoluble cancer drugs. *J. Am. Chem. Soc.*, **130**, 10876–10877.

74. Wallace, G.G., Spinks, G.M., Kane-Maguire, L.A.P., and Teasdale, P.R. (2003) *Conductive Electroactive Polymers: Intelligent Material Systems*, 2nd edn, CRC Press, Boca Raton.

75. Thompson, B.C., Moulton, S.E., Ding, J., Richardson, R., Cameron, A., O'Leary, S., Wallace, G.G., and Clark, G.M. (2006) Optimising the incorporation and release of a neurotrophic factor using conducting polypyrrole. *J. Control. Release*, **116**, 285–294.

76. Richardson, R.T., Thompson, B., Moulton, S., Newbold, C., Lum, M.G., Cameron, A., Wallace, G., Kapsa, R., Clark, G., and O'Leary,

S. (2007) The effect of polypyrrole with incorporated neurotrophin-3 on the promotion of neurite outgrowth from auditory neurons. *Biomaterials*, **28**, 513–523.

77. Quigley, A.F., Razal, J.M., Thompson, B.C., Moulton, S.E., Kita, M., Kennedy, E.L., Clark, G.M., Wallace, G.G., and Kapsa, R.M.I. (2009) A conducting-polymer platform with biodegradable fibers for stimulation and guidance of axonal growth. *Adv. Mater.*, **21**, 4393–4397.

78. Wong, Y.T., Dommel, N., Preston, P., Hallum, L.E., Lehmann, T., Lovell, N.H., and Suaning, G.J. (2007) Retinal neurostimulator for a multifocal vision prosthesis. *IEEE Trans. Neural Syst. Rehabil. Eng.*, **15**, 425–434.

79. Hodgson, A.J., Gilmore, K., Small, C., Wallace, G.G., MacKenzie, I.L., Aoki, T., and Ogata, N. (1994) Reactive supramolecular assemblies of mucopolysaccharide, polypyrrole and protein as controllable biocomposites for a new generation of 'intelligent biomaterials'. *Supramol. Sci.*, **1**, 77–83.

80. Kim, D.H., Richardson-Burns, S.M., Hendricks, J.L., Sequera, C., and Martin, D.C. (2007) Effect of immobilized nerve growth factor on conductive polymers: electrical properties and cellular response. *Adv. Funct. Mater.*, **17**, 79–86.

81. Innis, P.C., Moulton, S.E., and Wallace, G.G. (2007) in *Conjugated Polymers: Processing and Applications* (eds T.A. Skotheim and J.R. Reynolds), CRC Press, Boca Raton, pp. 11-11–12-12.

82. Garner, B., Hodgson, A.J., Wallace, G.G., and Underwood, P.A. (1999) Human endothelial cell attachment to and growth on polypyrrole-heparin is vitronectin dependent. *J. Mater. Sci. Mater. Med.*, **10**, 19–27.

83. Kotwal, A. and Schmidt, C.E. (2001) Electrical stimulation alters protein adsorption and nerve cell interactions with electrically conducting biomaterials. *Biomaterials*, **22**, 1055–1064.

84. Wallace, G.G. and Moulton, S.E. (2009) Organic bionics: molecules, materials and medical devices. *Chem. Aus.*, **76** (5), 3–8.

85. Cui, X., Lee, V.A., Raphael, Y., Wiler, J.A., Hetke, J.F., Anderson, D.J., and Martin, D.C. (2001) Surface modification of neural recording electrodes with conducting polymer/biomolecule blends. *J. Biomed. Mater. Res.*, **56**, 261–272.

86. Stauffer, W.R. and Cui, X.T. (2006) Polypyrrole doped with 2 peptide sequences from laminin. *Biomaterials*, **27**, 2405–2413.

87. Richardson-Burns, S.M., Hendricks, J.L., Foster, B., Povlich, L.K., Kim, D.H., and Martin, D.C. (2007) Polymerization of the conducting polymer poly(3,4-ethylenedioxythiophene) (PEDOT) around living neural cells. *Biomaterials*, **28**, 1539–1552.

88. Richardson-Burns, S.M., Hendricks, J.L., and Martin, D.C. (2007) Electrochemical polymerization of conducting polymers in living neural tissue. *J. Neural Eng.*, **4**, L6–L13.

89. Gilmore, K.J., Kita, M., Han, Y., Gelmi, A., Higgins, M.J., Moulton, S.E., Clark, G.M., Kapsa, R., and Wallace, G.G. (2009) Skeletal muscle cell proliferation and differentiation on polypyrrole substrates doped with extracellular matrix components. *Biomaterials*, **30**, 5292–5304.

90. George, P.M., Lyckman, A.W., LaVan, D.A., Hegde, A., Leung, Y., Avasare, R., Testa, C., Alexander, P.M., Langer, R., and Sur, M. (2005) Fabrication and biocompatibility of polypyrrole implants suitable for neural prosthetics. *Biomaterials*, **26**, 3511–3519.

91. Gelmi, A., Higgins, M.J., and Wallace, G.G. (2010) Physical surface and electromechanical properties of doped polypyrrole

biomaterials. *Biomaterials*, **31**, 1974–1983.

92. Schmidt, C.E., Shastri, V.R., Vacanti, J.P., and Langer, R. (1997) Stimulation of neurite outgrowth using an electrically conducting polymer. *Proc. Natl. Acad. Sci. U.S.A.*, **94**, 8948–8953.

93. Zong, X., Bien, H., Chung, C.Y., Yin, L., Fang, D., Hsiao, B.S., Chu, B., and Entcheva, E. (2005) Electrospun fine-textured scaffolds for heart tissue constructs. *Biomaterials*, **26**, 5330–5338.

94. Razal, J.M., Kita, M., Quigley, A.F., Kennedy, E., Moulton, S.E., Kapsa, R.M.I., Clark, G.M., and Wallace, G.G. (2009) Wet-Spun biodegradable fibers on conducting platforms: novel architectures for muscle regeneration. *Adv. Funct. Mater.*, **19**, 3381–3388.

95. Cheng, D., Xia, H., and Chan, H.S. (2004) Facile fabrication of AgCl@polypyrrole-chitosan core-shell nanoparticles and polymeric hollow nanospheres. *Langmuir*, **20**, 9909–9912.

96. Wan, Y., Yu, A., Wu, H., Wang, Z., and Wen, D. (2005) Porous-conductive chitosan scaffolds for tissue engineering II. in vitro and in vivo degradation. *J. Mater. Sci. Mater. Med.*, **16**, 1017–1028.

97. Breukers, R.D., Gilmore, K.J., Kita, M., Wagner, K.K., Higgins, M.J., Moulton, S.E., Clark, G.M., Officer, D.L., Kapsa, R.M.I., and Wallace, G.G. (2010) Creating conductive structures for cell growth: growth and alignment of myogenic cell types on polythiophenes. *J. Biomed. Mater. Res. A*, **95A**, 256–268.

98. De Giglio, E., Sabbatini, L., Colucci, S., and Zambonin, G. (2000) Synthesis, analytical characterization, and osteoblast adhesion properties on RGD-grafted polypyrrole coatings on titanium substrates. *J. Biomater. Sci. Polym. Ed.*, **11**, 1073–1083.

99. Yuanyuan, D., Jun, J., Shaofeng, Z., Yueling, Y., and Zhongyi, W. (2007) Preparation of polypyrrole coating on the pure titanium and its effects on osteoblast growth. *Rare Met. Mater. Eng.*, **36**, 91–95.

100. De Giglio, E., Cometa, S., Calvano, C.D., Sabbatini, L., Zambonin, P.G., Colucci, S., Benedetto, A.D., and Colaianni, G. (2007) A new titanium biofunctionalized interface based on poly(pyrrole-3-acetic acid) coating: proliferation of osteoblast-like cells and future perspectives. *J. Mater. Sci. Mater. Med.*, **18**, 1781–1789.

101. Serra Moreno, J., Panero, S., Materazzi, S., Martinelli, A., Sabbieti, M.G., Agas, D., and Materazzi, G. (2009) Polypyrrole-polysaccharide thin films characteristics: electrosynthesis and biological properties. *J. Biomed. Mater. Res. A*, **88**, 832–840.

102. Castano, H., O'Rear, E.A., McFetridge, P.S., and Sikavitsas, V.I. (2004) Polypyrrole thin films formed by admicellar polymerization support the osteogenic differentiation of mesenchymal stem cells. *Macromol. Biosci.*, **4**, 785–794.

103. Sirivisoot, S., Pareta, R.A., and Webster, T.J. (2009) Electrically-controlled penicillin/streptomycin release from nanostructured polypyrrole coated on titanium for orthopedic implants. *Solid State Phenom.*, **151**, 197–202.

104. Li, P., Song, W., Liu, L., Zhang, M., and Fan, Y. (2010) Effects of electrical and mechanical stimulation on mouse osteoblast like cells cultured on polypyrrole membrane. *Bone*, **47**, S424–S425.

105. Sirivisoot, S., Pareta, R., and Webster, T.J. (2011) Electrically controlled drug release from nanostructured polypyrrole coated on titanium. *Nanotechnology*, **22**, 085101.

106. Salto, C., Saindon, E., Bolin, M., Kanciurzewska, A., Fahlman, M., Jager, E.W., Tengvall, P., Arenas,

E., and Berggren, M. (2008) Control of neural stem cell adhesion and density by an electronic polymer surface switch. *Langmuir*, **24**, 14133–14138.

107. Zhang, L., Stauffer, W.R., Jane, E.P., Sammak, P.J., and Cui, X.T. (2010) Enhanced differentiation of embryonic and neural stem cells to neuronal fates on laminin peptides doped polypyrrole. *Macromol. Biosci.*, **10**, 1456–1464.

108. Peralta-Videa, J.R., Zhao, L., Lopez-Moreno, M.L., de la Rosa, G., Hong, J., and Gardea-Torresdey, J.L. (2011) Nanomaterials and the environment: a review for the biennium 2008–2010. *J. Hazard. Mater.*, **186**, 1–15.

109. Oberdorster, G., Oberdorster, E., and Oberdorster, J. (2005) Nanotoxicology: an emerging discipline evolving from studies of ultrafine particles. *Environ. Health Perspect.*, **113**, 823–839.

110. Lewinski, N., Colvin, V., and Drezek, R. (2008) Cytotoxicity of nanoparticles. *Small*, **4**, 26–49.

111. Hussain, M.A., Kabir, M.A., and Sood, A.K. (2009) On the cytotoxicity of carbon nanotubes. *Curr. Sci.*, **96**, 664–673.

112. Asplund, M., Thaning, E., Lundberg, J., Sandberg-Nordqvist, A.C., Kostyszyn, B., Inganas, O., and von Holst, H. (2009) Toxicity evaluation of PEDOT/biomolecular composites intended for neural communication electrodes. *Biomed. Mater.*, **4**, 045009.

113. Wang, X., Gu, X., Yuan, C., Chen, S., Zhang, P., Zhang, T., Yao, J.,

Chen, F., and Chen, G. (2004) Evaluation of biocompatibility of polypyrrole in vitro and in vivo. *J. Biomed. Mater. Res. A*, **68**, 411–422.

114. Green, R.A., Lovell, N.H., Wallace, G.G., and Poole-Warren, L.A. (2008) Conducting polymers for neural interfaces: challenges in developing an effective long-term implant. *Biomaterials*, **29**, 3393–3399.

115. Rutala, W.A., Weber, D.J., Weinstein, R.A., Siegel, J.D., Pearson, M.L., Chinn, R.Y.W., DeMaria, A., Lee, J.T., Scheckler, W.E., Stover, B.H. *et al.* (2008) *Guideline for Disinfection and Sterilization in Healthcare Facilities* (eds H.I.C.P.A. Committee), Department of Health and Human Services USA, Chapel Hill, NC.

116. Baveja, C.P. (2009) *Textbook of Microbiology*, 3rd edn, Arya Publications, New Delhi.

117. Ananthanarayan, R. and Paniker, C.K.J. (2009) *Textbook of Microbiology*, 8th edn, Orient Longman Private (Ltd), Hyderabad, India.

118. (2008) Trends in Radiation Sterilization of Health Care Products. Report. International Atomic Energy Agency, Vienna.

119. Moisan, M., Barbeau, J., Crevier, M.-C., Pelletier, J., Philip, N., and Saoudi, B. (2002) Plasma sterilization. Methods and mechanisms. *Pure Appl. Chem.*, **74**, 349–358.

120. (2007) Biosafety in Microbiological and Biomedical Laboratories (BMBL). Report. Centers for Disease Control and Prevention, USA: Office of Health and Safety.

5
Materials Processing/Device Fabrication

5.1
Introduction

The organic conductors discussed here must be endowed with appropriate structure to render the electrode–cellular interface highly effective. The electrode components must be integrated into a device comprising the source of electrical stimulation and, where appropriate, with some recording or electronic feedback system, as described in Chapter 1. In some situations, the formation of composite materials containing conducting polymers can enhance processability as well as other physical and chemical properties, and is thus worthy of consideration in the development of medical bionics.

Composite materials are those that consist of two or more constituent materials that do not completely merge to lose their identities despite amalgamating and subsequently imparting their characteristics in synergy to the overall material. In the form of composite materials, the properties, such as modulus, of the individual components can differ significantly from those of the individual constituents alone. The goal is to produce a composite material in which distinct properties and function(s) that were not observed individually in the constituent materials arise. Here, we refer to "organic bionic composites" as materials that utilize organic conducting polymers (OCPs) and/or carbon-based conductive nanomaterials (carbon nanotubes (CNTs), fullerenes, and graphene) and at least one other constituent, whether that is an inorganic, organic, or a biologically active species, that together give rise to new physical and chemical functions.

With the many different, in fact almost limitless, possible combinations of constituents, the invention of new classes of material with novel structures appears limited only by human creativity and/or desire. Rational strategies for producing composites readily allow novel materials to be selected for improved processability, superior or enhanced bulk properties, biological mimicry, multifunctionality, or specific targeting of biological function. From this perspective, this ongoing selection process to continually enhance the properties and function is a human-driven form of materials' evolution.

Organic Bionics, First Edition. Gordon G. Wallace, Simon E. Moulton, Robert M.I. Kapsa, and Michael J. Higgins.
© 2012 Wiley-VCH Verlag GmbH & Co. KGaA. Published 2012 by Wiley-VCH Verlag GmbH & Co. KGaA.

Traditional approaches to composite development generally involve choosing the desired materials classification, for example, a reinforced composite, and then a protocol for merging the constituent materials, which, for example, may involve factors such as the augmentation of strength or surface energy properties. This process is fundamentally akin to arithmetic procedures whereby the modes may involve the addition, averaging, division, and supplementation of the constituents' properties or function [1]. In many biomedical implants, modern engineering of a composite traditionally involves modulating the stiffness and strength of a material through a fiber-reinforced thermoplastic or thermosetting polymer matrix (e.g., a fiber-reinforced plastic). Different percentage volume contents of typically carbon fibers, glass fibers, or their particulate forms are incorporated into a wide range of polymeric materials (e.g., polyetheretherketone (PEEK), polymethylmethacrylate (PMMA)) to reinforce hip replacement screws and stems, orthopedic bone plates and dental implants [2].

For organic bionic composites, the complexity increases exponentially for the materials fabrication processes and chemistries required for the final composite structure. Initially, the bulk physical and surface properties (chemical composition, conductive, mechanical properties, etc.) of the composite material are of importance, as are their derivative micro- and nanoscale properties, including roughness, morphology, surface energy, wear resistance, brittleness, porosity, impedance, and so on. The inherent biocompatibility and toxicity of the composite must also be confirmed. It is at this point where traditional biomedical composites diverge from organic bionic composites, in that the latter require fine spatial and temporal control over a specific biological function. This necessitates the incorporation of biological constituents over the length scales of both cellular and molecular dimensions and within structures that are compatible with, or even mimic, the biological structures of the tissue into which they are placed. It is often desirable that the composite structure controllably releases biological constituents and then degrades itself over a specified time period. For example, it is envisaged that nerve regeneration conduits will require a discrete physical distribution of growth factors, inhibitory molecules, enzymes, and synthetic drugs, deposited onto biodegradable fibrous scaffolds [3, 4]. A further requirement is the need for an electronically conducting network to provide electrical stimulation to promote neurite outgrowth, as discussed in previous chapters. Such devices give rise to the need to develop innovative fabrication strategies to be realized in a timely manner.

The purpose of this chapter is thus to detail the different processing and fabrication strategies currently available in the field of medical bionics. The chapter first introduces:

1) Traditional processing methods such as blending and melting, with focus on using novel composite routes to improve the processing and handling of the conductive constituent; in most cases before further

modification and, to an extent, producing composites with enhanced electrical and physical properties (where possible).

2) Methods that can be used to fabricate organic conductors or composites containing them into practically useful forms include printing, extruding, spray drying, electrospinning/wet spinning and layer-by-layer, and dip-pen lithography techniques.

Throughout the course of this chapter, the issues and problems of ensuring practically useful medical bionic devices, including retaining the conductive properties and the activity and stability/structural integrity of the biological constituent, are considered.

5.2
Conducting Polymers

5.2.1
Blending

For most conventional polymer composite structures, simple mixing or blending of their constituents is used to achieve desired properties. Conventional polymers are by and large amenable to this approach since they are either soluble in common organic solvents or otherwise fusible, in that they melt before decomposing. This has generally not been the case with OCPs, as their strong van der Waals interchain attractive forces mitigate good solubility in organic solvents. They have, therefore, commonly been prepared in the form of intractable films, pellets, and powders, which often prove difficult to subsequently blend with other materials. The growing desire to use OCPs as constituent materials has seen significant developments in overcoming these solubility issues, primarily through modifications to the polymer backbone structure or addition of stabilizers. Outlined below are advances in the integration of OCPs into other polymer structures by either solution or melt processes.

5.2.2
Solution Processing

Polyaniline (PANi) is most amenable to solution processing. The emeraldine base (EB) form of PANi is soluble in selected solvents such as methyl pyrrolidinone [5] or strong acids [6, 7]. The solubility of the doped form can be induced by the use of appropriate "surfactant-like" molecules as dopants [8]. Camphor sulfonic acid (CSA) or dodecyl benzene sulfonic acids (DBSAs) have proved particularly useful in this regard. Once solubilized, these PANis can be cast into sheets or blended with other conventional polymer structures. For example, the presence of the surfactant counterion molecule facilitates the blending process [9] with polymers such as

1) polyethylene,
2) nylons,
3) polycarbonate (PC),
4) polystyrene (PS),
5) poly(vinylacetate), and
6) poly(vinylchloride).

When PANi/CSA is dissolved in *m*-cresol with PMMA, this can be spun cast to form optically transparent films [9]. Pron and coworkers [10] have blended PANi with cellulose acetate and cast from *m*-cresol to produce highly transparent and conductive ($\approx$1 S cm^{-1}) PANi. Addition of plasticizers (in particular poly(phosphoric acid), PPA) not only resulted in more flexible films but also lowered the percolation threshold for PANi to an amazing 0.05% (w/w) (Figure 5.1).

De Paoli and coworkers [11] used the more soluble poly(*o*-methoxyaniline) doped with *p*-toluene sulfonic acid to prepare a composite blend with poly(epichlorohydrin-*co*-ethylene oxide), but conductivities were relatively low (10^{-3} S cm^{-1}) even with a high 50% (w/w) conducting polymer loading. Others [12] have blended PANi doped with DBSA with polyvinyl alcohol (PVA) in water or with polyacrylate in organic solvents to obtain transparent conductive composites.

Solution processing of polypyrroles (PPys) and polythiophenes has been limited by these materials' lack of solubility. To overcome this, some work [13–15] has been focused on attaching alkyl groups to PPys or polythiophenes to increase solubility. This has had the desired effect of increasing solubility in common solvents such as CH$_2$Cl$_2$ or CHCl$_3$, in some cases to levels in excess of 300 g l^{-1} polymer. The solubility of polythiophenes has also been increased by attaching alkoxy groups to bithiophene monomers [16]. Other workers [17–19] have been concerned with producing pyrroles and thiophenes with alkyl-sulfonated chains attached to increase the water solubility of the polymer. Sulfonate groups have also been attached to PANi, either after [20] or before [21] polymerization in order to induce water solubility. The latter approach requires addition of a methoxy (electron donating group) to the aniline ring to counterbalance the effects of the electron withdrawing sulfonate group during polymerization (Figure 5.2; see chemical structure **1**).

However, by modifying the polymer backbone, solubility and hence processability are improved at the expense of the electrical (decreased conductivity) properties of the polymer. For the preparation of soluble poly(2-methoxyaniline-5-sulfonic acid) (PMAS) (Figure 5.2; see chemical structure **2**), a flow-through electrosynthesis approach has been devised [22, 23]. Subsequently, we developed a purification scheme that enables removal of low-molecular weight by-products that arise from the monomer oxidation

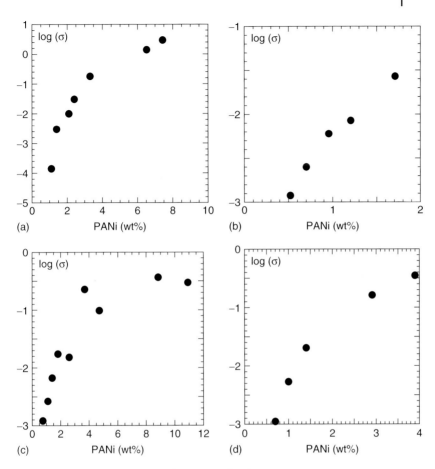

Figure 5.1 Conductivity in (S cm^{-1}) versus polyaniline content in PANi-plasticized CA composite films. PANi protonating acids: (a) camphor sulfonic acid (CSA), (b) poly(phosphoric acid) (PPA), (c) dibutyl phosphate (DBP), and (d) dioctyl phosphate (DOP). (Adapted with permission from Pron and coworkers [10].)

process. Removal of these by-products has a dramatic effect on the electronic properties of the water-soluble polymer obtained, increasing conductivity by approximately two orders of magnitude to 1 S cm^{-1} [24].

5.2.3
Colloidal-Assisted Processing

Given that conducting polymers with typically the most desired electrochemical and/or chemical properties have an inherent lack of melt or

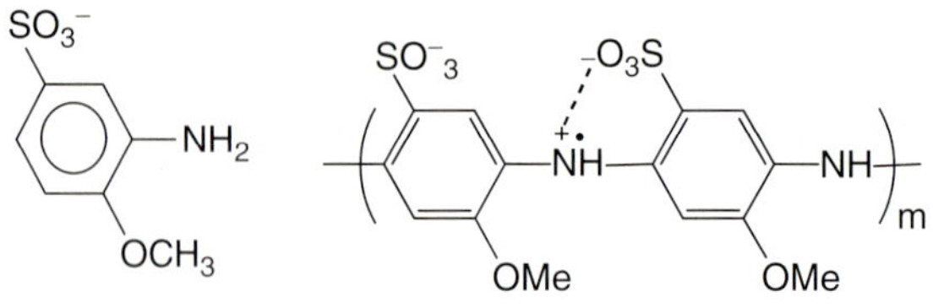

Figure 5.2 Chemical structure of 2-methoxyaniline-5-sulfonic acid (**1**). Chemical structure of poly(2-methoxyaniline-5-sulfonic acid) (PMAS) (**2**).

solvent processability, colloidal processing provides an attractive alternative. Conductive electroactive colloids of PPys or PANis have been prepared by carrying out the oxidative polymerization in the presence of a steric stabilizer [25], which physically adsorbs onto the growing polymer, preventing aggregation and macroscopic precipitation into its intractable forms. The process results in submicron-conducting polymer particles that are readily dispersed throughout other solutions for subsequent processing into composite structures (Figure 5.3).

Stabilizers such as PVA, polyvinyl pyrrolidone (PVP), and polyethylene oxide (PEO) have been used, producing conducting polymer particles with sizes down to the 10–100 nm range. While the conductivity of the colloidal dispersions are typically less than that of electrodeposited films (where the electrochemical film growth occurs directly on the substrate), significant conductivities of up to 10 S cm^{-1} have been achieved. Once formed, these colloids may be used for further processing. For example, they can be mixed with paints [26], which can be as conductive as those paint products obtained when carbon black is added. In some cases, larger conducting polymer particles have been mixed with other polymers. Bose and coworkers [27] incorporated chemically synthesized PPy into polyvinylchloride (PVC) or nafion. After casting, the materials were shown to be electroactive and demonstrated electrocatalytic properties. Our laboratories have implemented the electrochemical production of both PPy and PANi colloids under hydrodynamically controlled conditions via a flow-through cell [28, 29]. Although the various colloidal approaches have solved many of the solubility and processing issues, the commercial use of such colloidal products is currently limited because their conductivities are only comparable to their cheaper counterpart materials (e.g., carbon black). This said, the colloidal products are electroactive and presumably (at least the colloidal component) retain the unique chemical properties of a conducting polymer that may suit specific applications. Producing them as particles of nanometer scale also has the potential to introduce interesting materials properties.

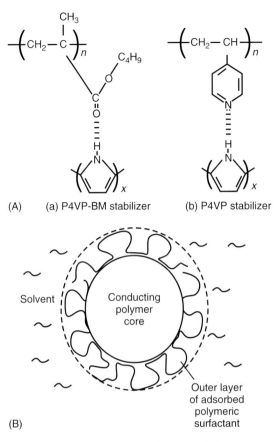

(A)   (a) P4VP-BM stabilizer   (b) P4VP stabilizer

(B)

Figure 5.3 (A) Representation of suggested adsorption modes for (a) poly(4-vinyl pyridine-*co*-butyl methacrylate) (P4VP-BM) and (b) poly(vinyl pyridine) (P4VP) stabilizers via hydrogen bonding onto polypyrrole. (B) Schematic representation of a single colloid particle of a conducting polymer. (Adapted with permission from Aldissi and Armes [25].)

5.2.4
Processing with Nanoparticles

5.2.4.1 **Inorganic Particles**
In an effort to improve the processability of nanoparticles, their mixing or blending with conventional nonconducting polymers in solution or melt form has been implemented. The situation is reversed for conducting polymer composites, as the initial role of the inorganic nanoparticles has been to facilitate the solubility/processability of the polymer constituent.

There is now interest in gaining value-added properties (e.g., thermal, catalytic, and chemical) from the vast array of inorganic particle types. For synthesizing such composites, commonly referred to as *conducting polymer "nanocomposites"* [30, 31], electrochemical or chemical polymerization of the monomer in the presence of the preformed inorganic particles appears to be preferred over simple mixing or blending.

Armes and coworkers [32] first initiated this approach by successfully encapsulating SiO_2 particles into host matrices of PANi and PPy. To achieve their objective of forming stable colloid dispersions, slow rates of chemical polymerization (using low concentrations of the monomer and chemical oxidant) were used to promote the growth of the conducting polymer on the particle(s) surface rather than in the bulk. Transmission electron microscopic (TEM) images have been used to describe a characteristic "raspberry" morphology for these composites (Figure 5.4). For 1 μm silica particles, relatively low conductivities of 2×10^{-5} S cm^{-1}(PANi/SiO$_2$) and 4×10^{-3} S cm^{-1} (PPy/SiO$_2$) were obtained, although higher values (4–7 S cm^{-1}) could be obtained by reducing the size of the particles. The charge-stabilization properties of the inorganic particles, held together at their surface by conducting polymer chains, were suggested to be responsible for the good colloidal stability of these composites.

Other researchers [33–35] have simultaneously polymerized both the conducting polymer and inorganic particles *in situ* to produce well-dispersed conducting polymer/nanoparticle composites and colloidal products, including Au/PEDOT, Pd/PPy, and Pd/PEDOT. The mechanism for composite formation in these colloidal products is a simultaneous oxidation–reduction reaction, where the monomer is oxidized by metal ions, while, at the same

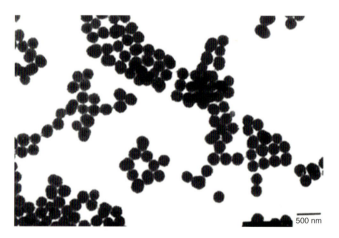

500 nm

Figure 5.4 TEM of a dilute dispersion of a PPy/silica colloidal composite. (Adapted with permission from Armes and coworkers [32].)

time, the metal ions are reduced to their neutral nanoparticle form. A Pd/PEDOT:PSS (colloidal) composite prepared using this process was used by Yin and Zhu [35] to fabricate thin-film devices with conductivities of $\approx 10^{-3}$ S cm^{-1}.

5.2.4.2 **Organic Nanoparticles**

The use of organic (nonconducting) nanoparticles has been utilized to a lesser extent and for a different purpose. Conventional polymers such as PS and polyurethane (PU) in the form of nanoparticles have primarily acted as constituents for template-assisted synthesis of the conducting polymer. Once incorporated, at which stage they form a conducting polymer nanocomposite, albeit short-lived, they are subsequently chemically removed to render void, honeycomb or shell structures. Some researchers have retained the nanoparticle constituent to fabricate core-shell structures <100 nm in size that show good dispersive, optical, and conductive properties [36].

While many of the above conducting polymers nanocomposites are primarily focused on battery, energy storage/conversion, catalytic, optical, and magnetic applications [30], the field is relatively new and they will find their way into medical bionics in due course. Already, metal-conducting polymer nanocomposites are presenting attractive properties for biosensing applications. For example, mercaptosuccinic-acid-capped gold nanoparticles (MSAGnp) in PANi can effectively dope the polymer and shift its electroactivity to neutral pH to enable electroactivity under physiological conditions [37] (Figure 5.5). PANi is normally redox active only in acidic conditions (pH < 4); thus, this provides an alternative route of introducing acidic groups to make feasible use of the PANi nanocomposite under biological conditions.

5.2.5
Melt Processing

For those conventional polymers that are insoluble or intractable, melt processing/blending of composites is a preferred option. This is an extremely common practice for thermoplastic polymers, particularly within industry. The process uses the application of high temperatures and shear stresses to transform polymer pellets into a viscous solution or "melt," at which any additives (e.g., conducting polymer, CNTs) can be introduced. The melt is then amenable to molding and extrusion processing to produce the desired structure. As with solution processing, there have been significant advances in melt processing of OCPs [38–40]. Again, most interest has been with PANi. The most successful approach has been to use specific dopants to induce melt processability. For example, various dopants are used by combining an acid group (e.g., phosphoric or sulfonic acid) with

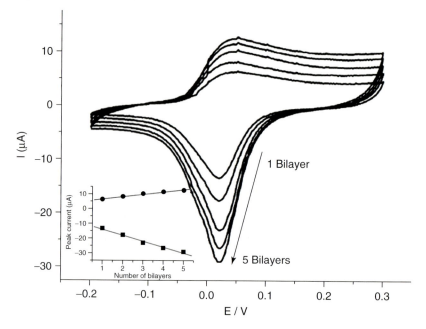

Figure 5.5 Cyclic voltammetry of PANi/(MSAGnp) composite recorded in 0.1 M phosphate buffer solution. (Adapted with permission from Tian and coworkers [37].)

hydrophobic segments. The hydrophobic segments cause a plasticization of the host PANi and reduce the strong interchain bonding caused by its aromaticity, H-bonding, and charge delocalization. Once suitably plasticized, the doped PANi is able to be fused to form free-standing films by hot pressing.

Paul and Pillai [39] used various phosphate ester dopants to induce melt processability in PANi (Figure 5.6a). Conductivities as high as 1.8 S cm^{-1} were reported for melt-pressed pellets of PANi/3-pentadecylpoly(phosphoric acid) (PDPPA) [40]. The doped polymers were found to be thermally stable up to 200 °C so that melt pressing at 160 °C did not induce any thermal degradation. Blends of the phosphate-ester-doped PANi were readily prepared with poly(vinyl chloride) by dry mixing at room temperature followed by hot pressing. The blends were found to give flexible films with a percolation threshold of 5% PANi content (by weight) and a conductivity of 2.5×10^{-2} S cm^{-1} for a blend having 30 wt% PANi/PDPPA in PVC. The same authors [40] have reported melt-processed PANi using sulfonic acid dopants (Figure 5.6b). These dopants also gave flexible free-standing films of doped PANi after hot pressing, with higher conductivities than the phosphate-ester-doped PANi (of up to 65 S cm^{-1} for doped systems prepared by an emulsion polymerization technique). The conductivity was dependent

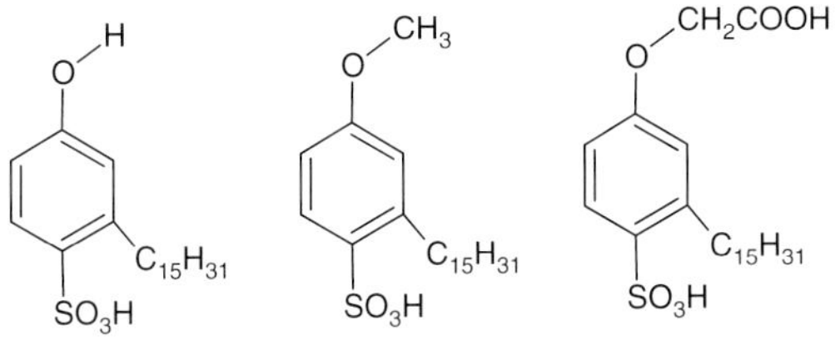

Figure 5.6 (a) Chemical structure of phosphoric acid derivative dopants. (b) Chemical structures of sulfonic acid dopants.

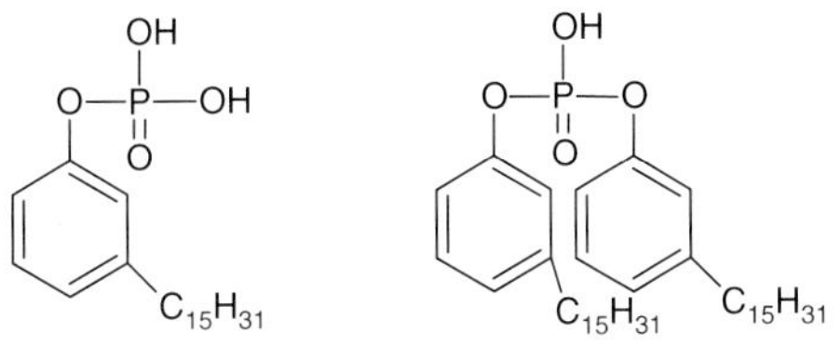

on the pressing temperature, reaching a maximum at around 140 °C. The decline in conductivity at higher pressing temperatures was probably due to thermal cross-linking of the PANi, reducing electron delocalization.

These demonstrations of melt processability for PANi provide an avenue for the convenient preparation of films and fibers of OCPs. The absence of solvent makes the melt processing route more attractive and environmentally friendly than alternative solvent processing. In addition, the melt processable OCPs are then easily incorporated into polymer blends using the well-developed processing techniques (extrusion and so on) widely used for thermoplastic polymers.

5.3
Carbon Nanotubes

Although advances in the synthesis and processing of CNTs are growing, these materials pose several challenges in the pursuit of CNT composites. Owing to their long aspect ratios, and the strong van der Waals interactions between them, CNTs easily aggregate to form bundles with poor dispersibility. The other issue is that transferring their remarkable molecular properties to macroscopic materials is hindered by their poor dispersion, adhesion, and alignment in the composite matrices. Achieving good CNT dispersions is critical for uniformly distributing their electrical, mechanical, and thermal properties throughout the composite. Even if a good dispersion is achieved, strong interfacial bonding is imperative to enable load transfer across the CNT–matrix interface. Several routes toward traditional mixing/blending via solution or melt processing have been developed to address these challenges.

5.3.1
Solution Processing

The simplest form of processing has been to mix a suspension of CNT material with thermoplastic or thermoset polymers, which involves producing an initial CNT dispersion in organic solvent using ultrasonication, followed by mixing with a polymer solution. Using a solution evaporation method, a multiwalled carbon nanotube (MWNT)/toluene dispersion was homogeneously mixed with PS dissolved in toluene to produce CNT/PS composite films [41]. The electrical conductivity of the CNT/PS composites was not measured in the study, although a 1 wt% of CNT loading (0.5% by volume) was shown to increase its mechanical strength by 36–42%. An additional increase in the break strength was due to the alignment and bridging of the CNT perpendicular to the crack propagation direction, thus providing closure stresses on the crack (Figure 5.7).

By mixing a CNT/nitromethane dispersion with a PMMA/nitromethane solution, casting of the mixed solution has produced 20–25 μm thick composite films [42]. At 15 wt% loadings, SWNT/PMMA (single-walled carbon nanotube) composite films had conductivities of 8.54–14.3 S cm^{-1} compared to MWNT/PMMA (0.15 S cm^{-1}) and vapor-grown carbon nanofibers/PMMA composites (0.06 S cm^{-1}). CNT loadings up to 50 wt% could produce values of 57.8 S cm^{-1}. We have recently dispersed CNT into a biomedical grade polymer, poly(styrene-β-isobutylene-β-styrene) (SIBS), previously used as drug delivery coatings on stents [43]. SIBS, which is an insulating block copolymer, was found to aid the dispersion of CNT, although the conductivity of its composite form was low (0.24 S cm^{-1}).

Solution-processing methods have been performed with different organic solvents and polymers [44, 45], typically culminating in the preparation of thin-film composites via solution casting and/or spin coating. In most cases, the solution processing can transform a highly resistive, conventional polymer ($\approx 10^{-16}$ S cm^{-1}) into a practical conductive composite (10^{-1}–10^{1} S cm^{-1}). The formation of stable CNT dispersions in aqueous solvent is also possible through their acid oxidation treatment, which creates electrostatically stabilizing polar, oxygen-containing surface groups [46]. Using this approach, composite films of CNT/PVA were prepared by careful mixing of their same solvent solutions, followed by casting and water evaporation [47] (Figure 5.8a). "Careful" mixing to prevent CNT aggregation was anecdotally described as allowing the CNT to become sufficiently covered, or sterically stabilized, by the polymer. The percolation (conductivity) threshold for these composites was between 5 and 10 wt%, with values reaching 10^{0} S cm^{-1} at 60 wt% of CNT loading (for 53 μm thick films) (Figure 5.8b).

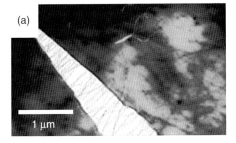

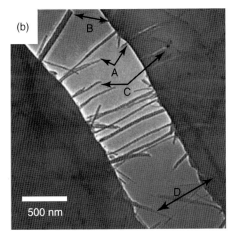

Figure 5.7 *In situ* TEM observation of crack nucleation and propagation in MWNT/PS films induced by thermal stress. (a) The crack propagates along weak CNT/PS interface or relatively low CNT density regions. (b) The cracks propagate along weak NT–PS interfaces or relatively low NT density regions (A and B). The MWNTs tend to align and bridge the crack wake then break and/or pull out of the matrix (C and D). (Adapted with permission from Qian and coworkers [41].)

Through the clever choice of the polymer matrix, solution processing has been exploited to remove impurities, in addition to dispersing the CNT as a composite [48]. The researchers showed that mixing a CNT powder with the conjugated polymer, poly(*m*-phenylene-vinylene-*co*-2,5-dioctyloxy-*p*-phenylene-vinylene) (P*m*PV), results in the selective interaction of the CNT and polymer. The CNT remained in solution, stabilized by the polymer, while the graphitic impurities sedimented out of solution.

5.3.2
Surfactant/Polymer-Assisted Processing

"Stabilizing" molecules, such as surfactants and polymers, have been used to facilitate the dispersion of CNT for their subsequent blending with polymers.

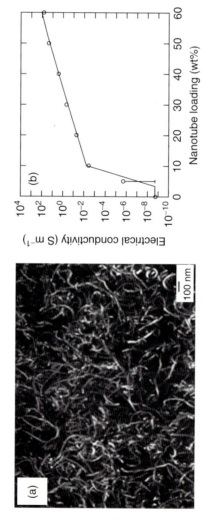

Figure 5.8 (a) SEM of a 50 wt% CNT/PVA composite. (b) Plot of electrical conductivities of CNT/PVA composites. The conductivity of the lowest loading fraction (5 wt%) represents the limit of what was measurable with the equipment used. (Adapted with permission from Shaffer and Windle [47].)

The nonionic surfactant, polyoxyethylene 8 lauryl ($C_{12}EO_8$), has been used to assist in the formation of CNT/epoxy composites, whereby the surfactant constituent's primary role was shown to be a dispersing agent [49]. For these CNT/epoxy composites, the CNTs interacted with the hydrophobic segments of the surfactant, leaving exposed hydrophilic regions to improve wetting behavior and/or subsequently enhance interfacial adhesion (via hydrogen bonding) between the constituents of the composite material. Sodium dodecylbenzene sulfonate (NaDDBS) and Triton X-100 are examples of other surfactants used to assist in solution processing of epoxy and PC, although the residual surfactant constituent can degrade the CNT conductive properties [50] or mechanical properties of the polymer matrix [51]. For amphiphilic surfactant stabilizers, palmatic acid was successfully used to efficiently disperse CNT in an epoxy matrix, although the electrical conductivity of this CNT/epoxy composite was low (10^{-6} S cm^{-1}) [52]. To avoid the use of organic solvents for blending conventional polymers, a microemulsion technique was implemented by mixing CNT with anionic (ammonium lauryl sulfate, sodium dodecyl sulfate (SDS)) and cationic (cetyltrimethylammonium bromide, CTAB) surfactants [53]. For this technique, polymerization of the styrene and isoprene monomers could be carried out *in situ* using oil-soluble initiators to produce CNT/PS and CNT/styrene–isoprene composites after sonication/coagulation of gray-bluish emulsions. DC electrical resistivity measurements in the study showed a substantial drop in values from 10^{16} Ωcm (unmodified polymer) down to 10^6 Ω cm for CNT/PS composites at 8 wt% CNT loading. The above use of *in situ* polymerization has mainly been limited to epoxy matrices, although CNT/PMMA composites have also been made [54]. Instead of surfactants, others [55] have added polymers such as poly(phenylene ethynylene)s (PPEs) to improve the initial CNT dispersion and mixing with another polymer. By incorporating the PPE polymer, PPE-CNT/PS and PPE-CNT/PS composites gave conductivities in the range of 10^0–10^2 S cm^{-1} at 7 wt% CNT loadings.

5.3.3
Chemical Modification

Chemical functionalization is a suitable method for controlling the interactive forces amongst CNTs, but importantly also with other constituents of the composite. To improve their dispersion qualities, CNTs were functionalized with 4-(10-hydroxy) decyl benzoate moieties followed by solution mixing with PS in toluene [56]. Incorporation of these CNT-4-(10-hydroxy) decyl benzoate into a PS matrix resulted in the formation of a percolated CNT network structure (at 1 vol% CNT), which was a desirable structure not achieved using nonfunctionalized CNT, and presumably due to the improved compatibility and CNT dispersion. Hill *et al.* [57] have also achieved intimate mixing with the polymer solution by chemically functionalizing the CNT with the same polymer type. The researchers functionalized CNT with a PS copolymer, poly(styrene-*co*-*p*-(4-(4'-vinylphenyl)-3-oxabutanol), PSV) and used solution processing (with PS at 5 vol% CNT) to produce 50 μm thick, transparent CNT–PS composite films. These CNT/PS composite films were shown to have a high optical quality, indicating the good dispersion of CNT throughout the PS matrix. Zhu *et al.* [58] attached alkylcarboxyl groups to SWNT via reactions with terminal diamines to enhance the reinforcement properties of CNT-epoxy composites. During the solution mixing, the terminal amino groups of the SWNT functionality readily acted with the epoxy and acted as a curing agent for the epoxy matrix. This ingenious approach caused the CNT to covalently integrate into the epoxy matrix and become part of the cross-linked structure rather than just a separate component, leading to 30–70% increases in the strength of the composite material with only small quantities (1–4 wt%) of alkylcarboxyl-CNT (Figure 5.9).

A majority of reports in this area only assess the mechanical attributes of the composites, which is not surprising given the wide belief that chemical functionalization disrupts the π-conjugation of CNT and hence their charge transport properties. However, some increases in conductivity have been reported.

5.3.4
Processing with Nanoparticles

Attaching nanoparticles to CNT is of particular relevance to bionics for electrode and biosensing applications. Metallic and metal oxide nanoparticles (e.g., platinum, gold, and copper nanoparticles), in particular, have fashioned this area as they are chemically stable, size controllable, and open to diverse chemical reactions at their surface to provide both chemical functionality and sensing capabilities. In early work, surface functional groups created by oxidation of CNT by H_2SO_4–HNO_3 were used as nucleation sites for the deposition of platinum clusters from a platinum salt solution [59]. The interaction of Pt^{2+} ions with surface carboxyl, carbonyl, or phenol groups to form nucleation precursors for subsequent platinum deposition via hydrogen reduction was proposed as the mechanism for these CNT–platinum composites. Jiang *et al.* [60] have used a simple method based on electrostatic adsorption to produce a nice uniform coating of gold nanoparticles, not clusters, on CNT. In this study, acid treatment was used to introduce carboxyl surface groups for the electrostatic attachment of the polyelectrolyte, poly(diallyldimethylammonium)chloride (PDADMAC), followed by anchoring the negatively charged particles to the PDADMAC (Figure 5.10).

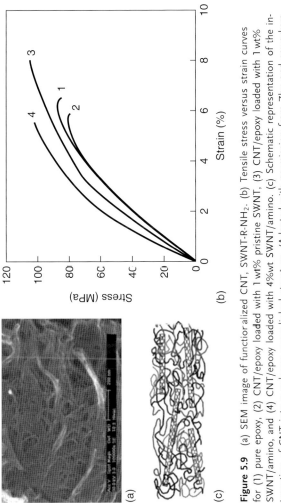

Figure 5.9 (a) SEM image of functionalized CNT, SWNT-R-NH$_2$. (b) Tensile stress versus strain curves for (1) pure epoxy, (2) CNT/epoxy loaded with 1 wt% pristine SWNT, (3) CNT/epoxy loaded with 1 wt% SWNT/amino, and (4) CNT/epoxy loaded with 4%wt SWNT/amino. (c) Schematic representation of the integration of CNT into a polymer cross-linked structure. (Adapted with permission from Zhu and coworkers [58].)

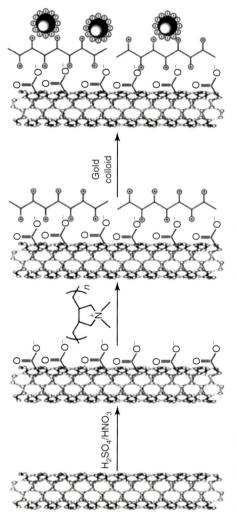

Figure 5.10 Schematic view of the process for anchoring gold nanoparticles to CN_x nanotubes. (Adapted with permission from Jiang and coworkers [60].)

In situ polymerization of gold nanoparticles, through the ingenious use of polyethyleneimine (PEI), has also provided a facile approach for fabricating CNT–Au composites [61]. In this case, the PEI acts as the functionalizing agent for the CNT, much like the polyelectrolytes above, and reducing agent for the formation of gold nanoparticles. To prepare CNT–platinum silicate composites as electrodes, although in the form of a paste, Yang *et al.* [62] have mixed acid-treated, dried CNT with a sol–gel platinum solution. The electroactivity of these CNT–platinum composite electrodes showed increased sensitivity, a decrease in the overvoltage of hydrogen peroxide production, and indicated a valuable use in conjugation with an enzyme such as glucose oxidase (GOx) for biosensing applications. For similar biosensing applications, nafion (perfluorsulfonated polymer) was used to initially solubilize CNT dispersions, but it also displayed strong interactions with Pt nanoparticles when their separate solutions were cast together to form modified electrodes [63]. Amperometric measurements showed that the CNT/nafion/Pt composite performed better than the individual constituents alone, with response times and detection limits of 3 s and 0.5 µM, respectively, for a standard GC/GOx system. These examples represent just some of the chemical bonding and intermolecular interaction pathways for immobilizing nanoparticles on CNT, as outlined in specific reviews on this topic [64].

5.3.5
Melt Processing

As nanomaterials, CNTs have a unique advantage in melt processing because they present less fiber breakage, maintain their high aspect ratio, and it is possible to align the CNTs during the process. CNT-nylon-6 (polycaprolactam) composites with different CNT loadings have been prepared by compression molding (250 °C and 150 bar pressure) [65]. At small 2 wt% CNT loadings, these composites showed dramatic improvements (>150% increase) in elastic modulus and yield strength (Figure 5.11).

Using melt mixing, Potschke *et al.* [66] have prepared CNT/PC composites with percolation thresholds between 1 and 1.5 wt% CNT loading. However, their DC conductivities were still low, with values of 10^{-8} S cm^{-1}. Further studies by these researchers showed that because of an increase in the melt viscosity and shear forces from adding CNT, the molecular weight of the PC matrix decreased, consequently negating the mechanical strength of the composite [67]. An alternative approach to adding the CNT has involved spraying aqueous CNT dispersions directly onto high-density polyethylene (HDPE) powders before melt mixing the CNT and the powder [68]. These CNT/HDPE composites had 4 wt% CNT loading percolation thresholds, with conductivities still relatively low (10^{-3} S cm^{-1}) even at higher 6 wt% loadings. Haggenmueller *et al.* [69] have incorporated alignment into the

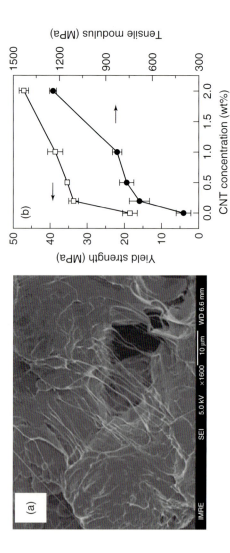

Figure 5.11 (a) SEM image of CNT/nylon composite at 0.5 wt% CNT loading. (b) Tensile stress (at yield) and tensile modulus as a function of weight percentage CNT loading. (Adapted with permission from Liu [65].)

Table 5.1 Summary and comparison of reinforcement of SWNT and MWNT composites fabricated by various processing methods.

Mechanical properties	Solution	Melt	Melt (fiber)	Epoxy	*In situ* poly	Functionalization
Mean reinforcement modulus (GPa)	309	23	128	231	430	157
Median reinforcement modulus (GPa)	128	11	38	18	60–150	115
Max reinforcement modulus (GPa) SWNT	112	68	530	94	960	305
Max reinforcement modulus (GPa) MWNT	1244	64	36	330	150	380
Max modulus (GPa)	7	4.5	9.8	4.5	167	29
Max strength (MPa)	348	80	1032	41	4200	107

Adapted with permission from Coleman and coworkers [71].

melt processing by subjecting solution-cast CNT/PMMA composite films to repeated cycles of hot pressing, followed by melt mixing in a flow apparatus. Owing to the greater CNT alignment in these CNT/PMMA composites, conductivity values of $0.11\,S\,cm^{-1}$ in the flow direction were slightly higher than the $0.07\,S\,cm^{-1}$ of the perpendicular direction. Andrews *et al.* [70] have also used extrusion processing of a melt to induce sample elongation and alignment of CNT in polypropylene matrix composites. While melt processing is simplistic and applicable to industry, conductivity values obtained from these composites appear low (e.g., $<10^{-2}\,S\,cm^{-1}$), as also their gains in mechanical properties, which generally underperform when compared to other processing methods (i.e., solution processing and chemical functionalization) [71] (Table 5.1).

5.4
Graphene

Apart from their remarkable theoretical properties, graphene has made a recent exciting entry into research on organic bionic composites because of the lessons learned over the past decade in processing their organic conductor "forerunners," especially the nanostructured carbons. Consequently, processing graphene into realizable commercial applications has developed in a very short time. In particular, the development of protocols for isolating and dispersing single graphene sheets suitable for processing has certainly paved the way for making graphene composites.

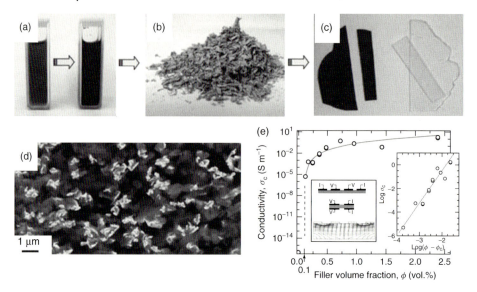

Figure 5.12 (a) Suspension of phenyl-isocyanate-treated graphite oxide and dissolved polystyrene in DMF before (left) and after (right) reduction in *N,N*-dimethylhydrazine. (b) Composite powder as obtained after coagulation in methanol. (c) Hot-pressed composite (0.12 vol% graphene) and pure polystyrene of the same 0.4 mm thickness and processed in the same way. (d) SEM image obtained from a fracture surface of composite samples. (e) Electrical conductivity of the graphene/PS composite as a function of volume filler fraction. (Adapted with permission from Stankovich and coworkers [72].)

5.4.1
Solution Processing

A general approach for the incorporation of graphene into polymer matrices was introduced by Stankovich *et al.* [72]. They used solution-phase mixing of exfoliated phenyl-isocyanate-treated graphite oxide sheets with PS, followed by chemical reduction, to disperse individual graphene sheets through the polymer matrix. Coagulation of the composite in methanol produced a powder form, which could then be hot pressed into 0.4 mm sheets. Figure 5.12 shows the different steps in the fabrication process and characterization of the graphene/PS composite's electrical properties. The graphene/PS film composites showed a percolation threshold of 0.1 wt%, a value three times lower than previously used 2D conductive fillers, and had absolute conductivities (10^{-2} S cm^{-1}) comparable to the best CNT–PS composites.

Others have used this same solution-processing method to fabricate thin-film field-effect transistors based on graphene/PS composites [73]. Solution processing of functionalized graphene sheets (FGSs) blended with PMMA has also achieved more thermomechanically stable composites compared to those prepared with expanded graphite and CNT fillers (Ramanathan *et al.*) [74]. For melt processing, Kim and Macosko [75] have obtained FGS from a thermally treated graphite oxide and then subsequently blended the graphene with PS via extrusion and molded the composite into various geometries. When comparing the graphene/PS composites to those made from flake graphite, the DC resistivity of the former decreased more sharply and at much lower concentrations (1.25 wt% FGS vs 6 wt% graphite loading). Further to this study, injection-molded graphene/PS composites had lower conductivities than annealed samples, attributed to an increase in graphene alignment (e.g., a less connective network). *In situ* polymerization in the presence of functionalized graphene (FG) nanosheet fillers has been used to prepare foliated graphite/nylon composites [76]. The percolation threshold for these composites occurred at 0.75 wt% FG nanosheets loading, with conductivity values reaching a steady state at 10^{-3} S cm^{-1}. A comprehensive study by Kim *et al.* [77] compared the performance of solvent-blending, *in situ* polymerization and melt compounding on the electrical properties of graphene/PU composites (Figure 5.13). Solvent processing and *in situ* polymerization both outperformed melt processing, as indicated by a significant decrease in their surface resistance at lower filler volume fractions. This trend agreed with TEM observations that showed the formation of undesirable graphene platelets in the melt-processed composites. Aqueous-based processing has also been feasible by mixing aqueous FGS dispersions, assisted by SDS, with an aqueous PEO to form cast film composites with conductivities of 10^{-2}–10^{-1} S cm^{-1} at 4 wt% loadings [78]. Alternative approaches to solution and melt processing include a latex-based method, incorporating a freeze drying step, which produced graphene/PS composites with conductivities of 0.15 S cm^{-1} at 1–2 wt% loading [79]. The use of FGS as a constituent embodies many of the earlier composite studies; however, it is noted that they are not the same as graphene. In general, most studies on graphene/polymer composites have been based on PS and PMMA matrices. Nylon, nafion, polypropylene, and epoxy resin matrices have also been used, with intercalation and subsequent polymerization of the monomer being the preferred route. The percolation thresholds for such graphene/polymer composites are typically <2.5 wt% loading [80]. Constituents, other than conventional polymers, range from TiO_2, Al_2O_3, Au-nanoparticles, Si-nanoparticles, conducting polymer, and CNT, although these are not presented in detail here.

5.5
Composites with Conventional Polymers–a Medical Focus

For conducting polymers, nanoparticulate PPy polymerized within a processable silicon elastomer host polymer to fabricate electrodes for neural

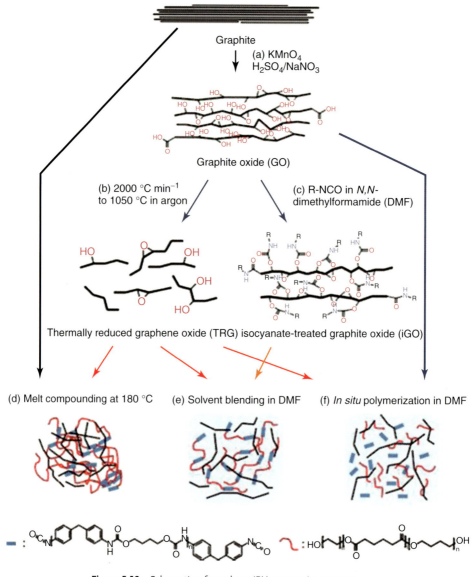

Figure 5.13 Schematic of graphene/PU composite preparation routes. Black lines represent graphitic reinforcements. PU hard and soft segments are denoted by short blue and thin red curved lines. (Adapted with permission from Kim and coworkers [77].)

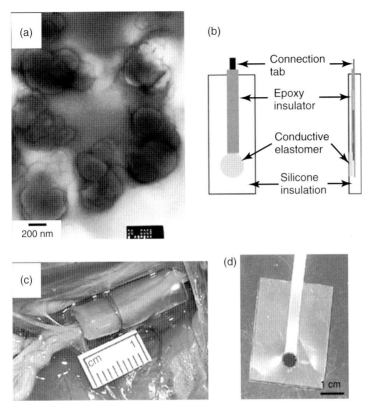

Figure 5.14 (a) TEM of chemically polymerized PPy in silicone elastomer. (b) Schematic of electrode fabricated from patterned layers of conductive elastomeric nanocomposite and insulating silicone rubber. (c) *In vivo* testing of single-contact prototype cuff electrode fabricated from the conductive elastomeric electrode prototype cuff electrode implanted on the cat sciatic nerve. (d) Photograph of conductive elastomeric electrode. (Adapted with permission from Keohan and coworkers [81].)

stimulation was studied both *in vitro* and *in vivo* [81]. Assessing the device performance of single prototype nerve cuff electrodes, following acute implantation, demonstrated the feasibility of using these composite materials for peripheral nerve stimulation (Figure 5.14). The study highlighted that matching the mechanical properties of the electrode to the underlying nerve tissue was important for future work.

Zhang *et al.* [82] have coated PPy onto polyester fabrics for vascular prostheses applications. The PPy was deposited on woven poly(ethylene terephthalate) (PET) fabric via vapor deposition subsequent to covalent bonding of pyrrole precursor moieties on the PET. The PPy/PET composites were shown to be promising as blood-contacting materials in *in vivo* studies and have further scope because of their conductive properties. Similar textile-like organic conducting composites formed by chemical polymerization of PPy onto silk fabrics with encouraging biocompatible and conductive properties were also proposed for innovative biomedical applications [83].

Microcatheter structures, for potential use as biomedical devices, have been fabricated using conventional processing of CNT/polymer composites [84] (Figure 5.15). For their fabrication, Nylon-6 and CNT (10 wt%) was solution mixed, dried, and then fed into a twin-screw extruder to produce microcatheters with inner and outer diameters of 0.46 and 0.53 mm, respectively. The microcatheters had an opaque black appearance and observations of the composite structures in cross section, after fracturing, indicated a uniform dispersion of the CNT. After a period of implantation into animal arteries, their lumen exhibited highly reduced thrombosis formation and low blood coagulation compared to pristine polymer microcatheters. Biocompatibility studies also showed that they were relatively inert to the surrounding tissue. The usefulness of their incorporated electronic properties was mentioned, although no specific applications were addressed.

5.6
3-D Structured Materials and Device Fabrication

5.6.1
Fiber Spinning Technologies

The assembly of fibers through weaving and other processes are the basis of the mass production of textiles, carpets, and even paper products. Utilizing these well-developed technologies for the construction of 3D structures, organic bionic composites, and related devices are extremely attractive prospects for medical bionics applications. For instance, it is in the emerging field of electronic textiles that fiber spinning and organic conducting technologies merge. An electronic textile contains electronic components that are seamlessly integrated into a conventional fabric structure. The seamless integration of computers and mobile phone technology built into wearables will have a dramatic impact on the field of bionics. The addition of conductive fibers to conventional fabrics to impart energy storage (batteries and capacitors) or energy conversion (photovoltaic and thermal energy harvesting) capabilities will also be extremely useful for powering medical bionic devices. Another application is in the area of biomedical monitoring. Here, a number of sensor fibers are built into garments and the sensor responses recorded, and in some scenarios automatically transmitted to health service providers. Already, solution-processable conducting polymers are being used to develop fiber products that may find use in the field of medical bionics.

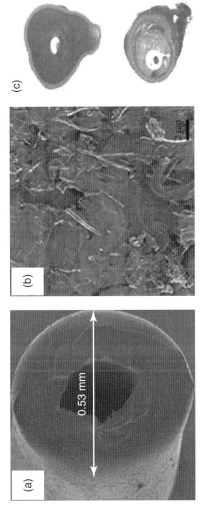

Figure 5.15 (a) SEM image of microcatheter composite. (b) Higher magnification SEM image of composite cross section showing uniform dispersion of CNT. (c) Cross-sectional views of pure nylon-derived (top) and composite microcatheter (bottom) placed into artery. (Adapted with permission from Koyama and coworkers [84].)

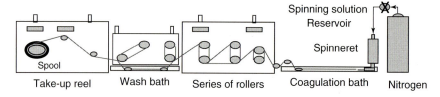

Figure 5.16 Spinning setup used to fabricate wet-spun polymer-based fibers.

5.6.2
Wet Spinning

The setup used for wet spinning of fibers is shown in Figure 5.16. A feed solution of appropriate viscosity is injected at a controlled rate (usually $1.8\text{--}8.5\ \text{ml}\ \text{h}^{-1}$) into a coagulation bath [85]. The fiber forms spontaneously as the feed solution is precipitated under these controlled hydrodynamic conditions. Some examples of feed solutions and coagulation baths used are given in Table 5.2. Many of the fiber constituents consist of biopolymers (e.g., chitosan, alginate, and DNA) to enhance biocompatibility or impart biological function. As discussed in the following sections, the feed solutions can also be blended with organic conductors (e.g., CNT, conducting polymers) to create composite, conducting fiber structures.

5.6.3
Conducting Polymers

A number of workers have shown that it is necessary to synthesize PANi to high molecular weights to successfully prepare fibers with adequate mechanical properties. Laughlin and Monkman [90] showed that a molecular weight of $130\ 000\ \text{g}\ \text{mol}^{-1}$ is sufficient, whereas Mattes and coworkers [91] have investigated the effect of molecular weight in the range $100\ 000\text{--}300\ 000\ \text{g}\ \text{mol}^{-1}$. In general, more material can be drawn and the tensile strength of the fiber increased when a higher molecular weight is used. Further work by Mattes and coworkers [92] has produced PANi-EB fibers (Figure 5.17). The as-wet-spun fibers had a modulus of $0.54\ \text{GPa}$, tensile strength of $15\ \text{MPa}$, and elongation at break of 9%, which were altered by drawing ($4\times$) to $1.85\ \text{GPa}$, $63\ \text{MPa}$, and 6%, respectively.

Direct preparation of PANi emeraldine salt (ES) fibers is also possible. Monkman and coworkers [93] have reported ES fibers with a modulus of $2\ \text{GPa}$ and tensile strength of $97\ \text{MPa}$ with a conductivity of $600\ \text{S}\ \text{cm}^{-1}$. PANi-ES/CNT composite fibers have been prepared in a one-step spinning process to enhance the PANi fiber properties [94]. The pristine PANi-ES drawn fibers had a modulus of $3.4\ \text{GPa}$, tensile strength of $170\ \text{MPa}$, and elongation at break of 9%. With the addition of 0.76% (w/w) CNT,

Table 5.2 Examples of feed solutions and coagulation baths.

Feed solution	Coagulation bath	Fiber produced	Reference
PLGA (75 : 25 and 85 : 15)	Isopropanol	PLGA	[85]
Sodium alginate	HCl (0.2 M) and/or CaCl$_2$ (1–3% w/v)	Alginate	[86]
Chitosan	Ethanol/Ca(OH)$_2$	Chitosan	[87]
SWNT/DNA	Polyvinyl alcohol	SWNT-DNA-PVA	[88]
SWNT/chitosan dispersion	Chondroitin sulfate (CS) heparin	SWNT-Chit-CS SWNT-Chit-Hep	[89]
SWNT/DNA dispersion	Chitosan	SWNT-DNA-Chit	[89]
SWNT/hyaluronic acid (HA)	Chitosan	SWNT-HA-Chit	[89]

these properties were altered to 7.3 GPa, 255 MPa, and 4%, respectively. In addition, the conductivity increased from 500 to 700 S cm^{-1} with the addition of the CNT. The composite fibers also displayed a high degree of flexibility and twistability (Figure 5.18). In comparison, PANi-ES fibers prepared by first wet spinning leucoemeraldine and then posttreating to produce the ES showed inferior mechanical and electrical properties, with modulus 3.5 GPa, tensile strength 45 MPa, elongation at break 4.6%, and conductivity 96 S cm^{-1} [95].

Wet spinning has been used to produce long lengths of polythiophene fibers. For example, poly(3,4-ethylenedioxythiophene) (PEDOT) fibers have been produced by wet spinning [96] an aqueous solution of PEDOT:PSS into an acetone containing coagulation bath. The wet-spun fibers resulted in good conductivities of ≈1.0 S cm^{-1}. Smooth and straight 5 μm diameter PEDOT:PSS fibers with higher conductivities of 11–74 S cm^{-1} have been fabricated from commercial-based solutions. In this work, the researchers attempted to enhance conductivity by wet spinning the microfibers from PEDOT:PSS dispersions containing ethylene glycol (EG); however, this failed because of the EG dissolving in the acetone coagulation bath, although subsequent dip coating of the PEDOT:PSS fibers in EG increased the conductivities up to 195–467 S cm^{-1}[97].

Recently, PPy has been rendered soluble through the use of di-(2-ethylhexyl sulfosuccinate) (DEHS) as the dopant (Figure 5.19, see chemical structure) [98]. The viscosity and surface tension of solutions containing PPy/DEHS (**1**) are such that they are amenable to wet spinning, using dichloroacetic acid (DCAA) as solvent and 40% dimethylformamide (DMF) as coagulant. PPy fibers with good mechanical (ultimate tensile strength = 25 MPa, elastic modulus 1.5 GPa) and electronic (3 S cm^{-1})

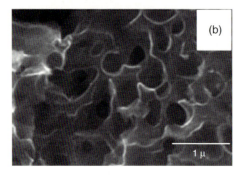

Figure 5.17 (a) SEM micrograph of a cross section of a PANi-EB fiber and (b) zoomed-in image of the cross section. (Adapted with permission from Mattes and coworkers [92].)

properties can be produced. An alternative approach to spinning PPy fibers is the use of a reactive wet-spinning protocol [99, 100]. This involves formation of the PPy during the wet-spinning process.

5.6.4
Carbon Nanotubes and Graphene

Pristine CNT fibers can be produced using wet spinning by using a co-agulating flow composed of ethanol/glycerol or ethanol/glycol mixtures as well as aqueous-based coagulating agents (e.g., aluminum nitrate solution) [101]. By washing the fibers with water/ethanol, the coagulating agents can be removed to leave behind pure CNT fibers. For composite fibers, SWNT/poly(p-phenylene benzobisoxazole) (PBO) in poly(phosphoric acid) (PPA) has been wet spun into fibers using a dry-wet spinning procedure [102]. It was shown that the tensile strength of the CNT/PBO fiber composite containing 10 wt% CNT was about 50% higher than that of the pristine PBO containing no CNT.

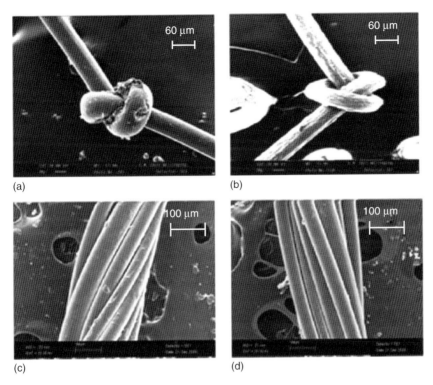

(a)

(b)

(c)

(d)

Figure 5.18 SEM images of knotted (a,b) and twisted (c,d) composite fibers (a) PANi fibers. (b) PANi/CNT composite fibers. (Adapted with permission from Mottaghitalab and coworkers [95].)

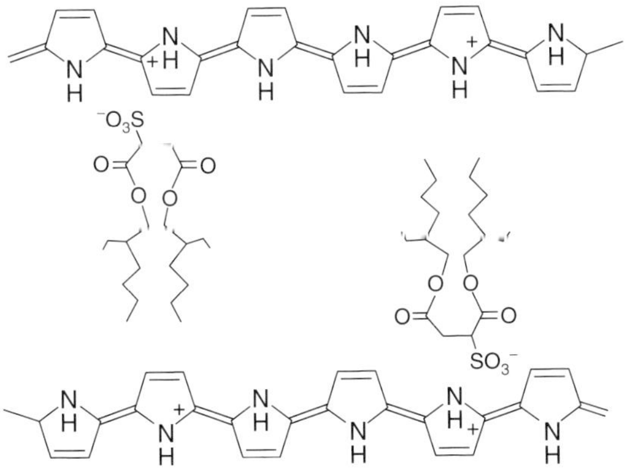

Figure 5.19 Schematic of interaction between Na$^+$DEHS$^-$ and oxidized polypyrrole.

We have also produced a body of work on wet-spun composite fibers using CNT as reinforcement additives for artificial muscle and biomedical applications. Through noncovalent functionalization, Razal *et al.* [103] were able to process CNT dispersions into long lengths of fiber using a wet-spinning approach (Figure 5.20). The fibers produced were mechanically strong and possessed conductivities that were impressive for composite materials; in addition, these fibers supported the growth of fibroblast cells with no apparent cytotoxic effects.

In other work, a CNT/chitosan composite fiber containing the drug dexamethasone was produced and controlled release under electrical stimulation was demonstrated. The production of microscaled drug delivery systems is appealing for biomedical applications, where precise targeted delivery is required. The incorporation of CNT into polymer-based wet-spun fibers has also produced mechanically strong fibers compared to their polymer-only fiber counterparts. Lynam *et al.* [89] utilized the charge properties of biological molecules to produce fibers composed of CNTs and molecules such as hyaluronic acid (HA), chitosan, and chondroitin sulfate (CS). The produced fibers exhibited good mechanical and electrical properties. More recently, we have incorporated chitosan and other biopolymers into wet-spun fibers comprising CNT to form composites, including chitosan/PANi/CNT fibers [104] and gellike CNT/chitosan or CNT/gellan gum fibers [105, 106] (Figure 5.21).

Specifically for medical bionic applications, fibrous structures are of particular interest, as electrically excitable cells are often arranged in fibrous aligned structures such as nerve and muscle tissue. Therefore, structures such as fibers are desirable as scaffolds for cell growth. Work in our laboratories has targeted the use of wet spinning to prepare biodegradable fibers, intimately integrated with conducting polymer structures, that have proved to be excellent scaffolds for supporting nerve and muscle growth [85, 107].

In contrast to conducting polymer- and CNT-based fibers, it appears that little or no work has been reported so far on the wet spinning of graphene fibers.

5.6.5
Electrospinning

Electrospinning enables rapid formation of long polymer fibers with diameters ranging from 100 nm to 2 μm [108–110]. The method involves dissolving the polymer in a suitable solvent and then applying a large voltage difference between a metal capillary containing the solution (e.g., a syringe) and a target (as illustrated in Figure 5.22). The target may be a metal foil on which the nanofibers deposit. The applied potential on the syringe needle creates a Taylor cone within the polymer droplet as the charged

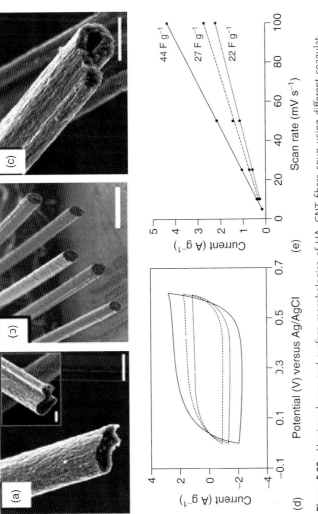

Figure 5.20 Varying shapes and surface morphologies of HA–CNT fibers spun using different coagulating media. (a) 0.3 M HNO₃ (scale bar 20 μm), (b) 5% (w/v) CaCl₂ in 70% (v/v) ethanol solution (scale bar 200 μm), and (c) ethanol (scale bar 30 μm). Inset in (a) shows similar fiber composition to fibers in (b), but they were spun using different spinning conditions (scale bar 20 μm). (d) Electrochemical behavior of HA–CNT fibers spun in 0.3 M HNO₃ (solid line), ethanol (dashed line), and 5 wt% CaCl₂ in 70 vol% methanol (dotted line). Cyclic voltammograms in 0.2 M PBS (pH 7.4) at 50 mV s⁻¹. (e) Plot of current versus scan rate, showing specific capacitances of fibers. (Adapted with permission from Razal and coworkers [103].)

Figure 5.21 (a) SEM image of CNT/gellan gum composite fiber and (b) zoomed-in image of the cross section. (Adapted with permission from Granero and coworkers [105, 106].)

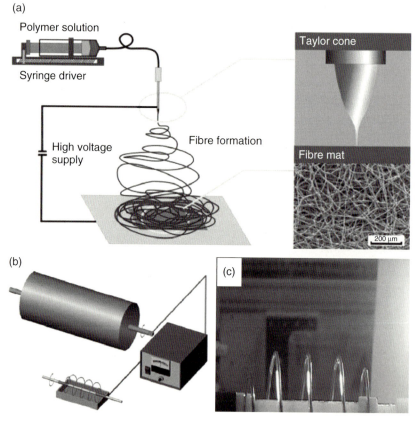

Figure 5.22 (a) Schematic diagram of electrospinning. (b) Needleless electrospinning design and (c) in operation.

polymer chains repel each other and the solvent evaporates. The fibers formed are collected on a grounded target as an entangled nanofibrous mat. Modifications to the electrospinning method have also been developed in order to align the fibers during the electrospinning process [109]. This approach has been used to produce aligned conducting polymer fibers [110]. Considering the dimensions of the wet-spun fibers and the types of available topographies, they are well suited to presenting topographical cues for directional growth in a medical bionic scaffold [111]. More recent advances in the electrospinning design have seen the emergence of needleless electrospinning (Figure 5.22b,c). This system uses a metal coil electrode rather than a single needle electrode resulting in large-scale electrospinning.

Electrospinning has been used to produce PPy/DBSA fibers in the form of a nonwoven mat that gave conductivities of $\approx 0.5\,\mathrm{S\,cm^{-1}}$ following compression [112]. An alternative approach involves electrospinning the fibers from a chemical oxidant/polymer-mixed solution followed by exposure of the fibers to pyrrole vapor [113]. This approach has resulted in PPy/PEO composite fibers of ≈ 100 nm in diameter and with conductivities of 10^{-3} $\mathrm{S\,cm^{-1}}$ for the fiber mats produced. The vapor-phase polymerization approach has been used to produce PPy/SIBS nanofibers as platforms for cell culturing [3, 114] (Figure 5.23). To increase the spinnability and control the conductivity of the PPy and other conducting polymer materials, they are often blended with PEO as this high-molecular-weight polymer is reported to increase chain entanglements [115]. The incorporation of insulating PEO does compromise conductivity.

The electrospinning of polythiophenes has been previously reported [116–121]. Electrospun PEDOT fibers have also been fabricated [122],

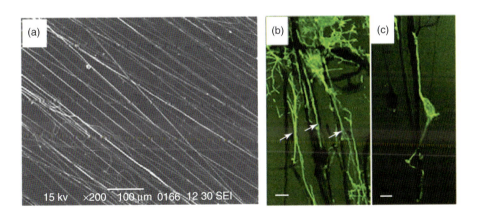

Figure 5.23 Figure 5.23 SEM images of aligned PPy/SIBS composite fibers (a). Fluorescence micrograph showing aligned neurites of nerve cells (b, arrows) on the fibers (dark lines). Scale bars in (b),(c) represent 10 μm. (Adapted with permission from Liu and coworkers [114].)

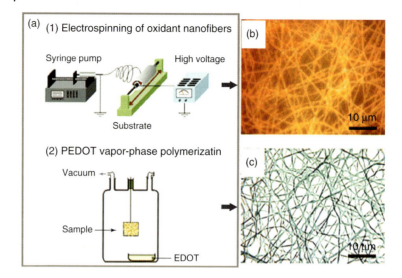

Figure 5.24 Scheme of the two-step nanofibers production (a). Optical micrographs of the resulting nanofibers, (b) oxidant fibers, and (c) PEDOT fibers. (Adapted with permission from Laforgue and Robitaille [123].)

producing nanofiber mats with conductivity as high as 7.5 S cm^{-1}. Numerous other routes to incorporating a conductive PEDOT coating onto electrospun fibers include the use of vapor-phase polymerization [123], electrochemical deposition [124], and dip coating [125]. The highest conductivity value (60 S cm^{-1}) claimed for polymer nanofibers to date is for PEDOT fibers produced via an electrospun polymer host containing oxidant and subsequent vapor-phase polymerization [123] (Figure 5.24).

The family of thiophenes represent an interesting material because of their unique flexibility for tailoring of polymer properties as a result of the ease of functionalization of the parent monomer; for example, the electrospun fiber preparation from ester-functionalized organic soluble polythiophenes (poly-octanoic acid 2-thiophen-3-yl-ethyl ester (POTE)) and post-polymerization hydrolysis of the ester linkages (Figure 5.25a,b). Such versatility in the functionality can be built on to improve the compatibility of the fibers with the biological tissue of interest. Electrospun functionalized polythiophene fibers have been shown to support the growth and differentiation of muscle cells [126] (Figure 5.26c,d) with aligned structures guiding the alignment of developing muscle fibers.

Several publications have reviewed the use of electrospinning to produce nanofibrous structures containing CNTs [127–129]. Electrospun bio-nanowebs based on CNT/SIBS composites [130] and CNT/PEO fibers have been produced [131].

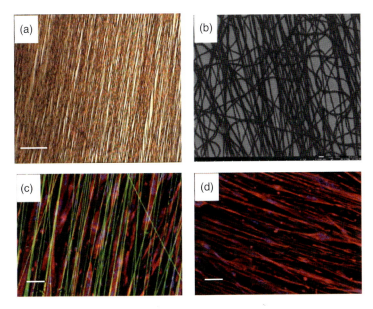

Figure 5.25 Aligned POTE fibers: (a) microscope image, scale bar represents 20 μm; (b) SEM image, scale bar represents 10 μm. Fluorescence images of differentiated primary myotubes, (c) aligned along medium density fibers of POTE or (d) poly(HET). (Reproduced from unpublished work: R.D. Breukers, PhD Thesis, University of Wollongong 2011.)

5.6.6
Printing Technologies

Printing is also an old art that is now seen in a new light for fabrication of advanced products, including electronic components. Solution-processable OCPs and nanostructured carbons can be developed as formulations that are inherently suitable for fabrication using printing methods. This is already leading to low-cost sensors and cheap, disposable electronics. It would appear that the manufacturing protocol set up for printed electronics provides a flexible platform to develop manufacturing protocols for printed medical bionic devices.

Of the printing strategies available, inkjet printing and extrusion printing provide the resolution and flexibility required to print medical bionic devices. Inkjet printing technology provides a resolution of tens of micron and the ability to readily produce complex patterns in two dimensions. Extrusion printing provides a route to three-dimensional structures with tens of microns resolution. Dip-pen nanolithography (DPN) also deserves mention as a printing technology of the future that is capable of providing nanometer resolution.

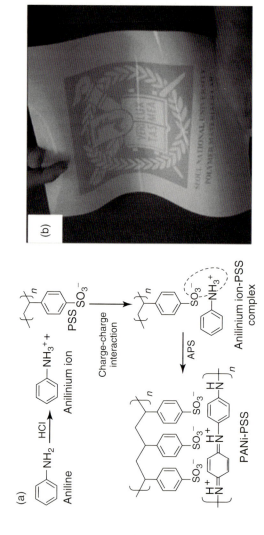

Figure 5.26 (a) Synthetic procedure for the production of water-dispersible PANi/PSS. (b) Photopaper image of the emblem of Seoul National University printed with an inkjet printer using water-dispersible PANi/PSS. (Adapted with permission from Jang and coworkers [137].)

5.6.7
Inkjet Printing

Inkjet printing is a digital, noncontact printing technology wherein the ink is transferred and patterned directly onto the substrate from a nozzle or a printhead. This allows a great deal of flexibility at the research and development stage. Printed patterns can be evaluated and changed rapidly and at minimal cost using simple CAD software. Inkjet printing can also be performed over a large area with high throughput by combining many printheads into large linear arrays. However, to achieve patterning, the substrate or the printhead, or a combination of both, must be moved. Typically, for small-scale developmental work, printers with a movable XY printhead are used. For large-scale production, single-pass linear arrays of inkjet printheads are employed. Feature sizes as small as 10–20 µm are now being reported.

Inkjet technologies can be classified as continuous and drop-on demand. Continuous inkjet technology generates a constant stream of small ink droplets, which are charged according to the image and controlled electronically. The charged droplets are deflected by a subsequent electric field, while the uncharged ones flow onto the substrate.

Drop-on-demand inkjet technology, on the other hand, produces a droplet only if it is required by the image. Thermal inkjet and piezo inkjet printing are known as *the most important drop-on-demand technologies*. Thermal inkjet (bubble jet) generates the drops by heating and localized vaporization of the liquid in a jet chamber. However, thermal inkjet printers have some limitations. Usually, only water can be used as the solvent for the ink because of its vapor point, and this imposes restrictions on printing materials that only dissolve in organic solvents to form nonaqueous thermal inks. With a piezo inkjet, the ink drop is formed and catapulted out of the nozzle by mechanically deforming the jet chamber, an action resulting from an electronic signal and the piezoelectric properties of the chamber wall. Therefore, inks based on either organic or aqueous solvents are suitable for piezoelectric inkjet printers. For both printer types, nozzle sizes are typically 20–30 µm in diameter, droplet volumes produced are 10 20 pl, and final drop diameters are in the range of 30–40 µm. Smaller nozzles allow for smaller droplets to be produced and higher print resolution, with drop volumes of 1 pl possible.

Crucial aspects of inkjet printing technology are the ink and its physical properties, in particular, the viscosity and surface tension. The viscosity should be low enough to allow the channel to be refilled in about 100 µs, and surface tension should be such that the ink is held in the nozzle without dripping [132]. In general, a typical viscosity for an inkjet printable ink is below 20 mPa s, and, in practice, a surface tension in the range from 28 to 350 mN m^{-1} [133] is acceptable. Particle size and distribution of particles in the ink are also important factors for printing. The recommended particle size is dependent on the printhead used but is typically below 1 μm and ideally <200 nm. To maintain a small particle size, it is important that agglomeration does not occur over time (i.e., any ink dispersion must be stable). When the particle size becomes of the order of 1 μm, printability problems may occur.

Aqueous PANi nanodispersions doped with DBSA have been produced [134] with a uniform particle size of approximately 80 nm in the doped conductive form. These nanodispersions have been found to be highly stable over long periods of time (at least several months). A rheological study indicated that PANi-DBSA nanodispersions could be prepared as inks with viscosities suitable for piezoelectric inkjet printing (2–12 mPa s). Surface tensions of the nanodispersions (27–30 mN m^{-1}) were also similar to that of a commercial inkjet printing ink. High-quality images were generated by inkjet printing on paper. It was shown that both electrochemically deposited [135] and inkjet-printed films formed from the nanoparticles were comparable with electrochemically polymerized films in terms of conductivity and electroactivity [136].

PANi–poly(4-styrenesulfonate) (PSS) nanoparticles (average diameter = 28 nm) have also been used to form aqueous dispersions with physical properties that make them amenable to inkjet printing [137] (Figure 5.26). The viscosity and surface tension were found to be about 4 mPa s and about 63 mN m^{-1}, respectively, making them suitable for inkjet printing. Inkjet-printed sensors on photo paper (Figure 5.26) showed superior sensitivity and more rapid response time relative to conventional photolithography printed PANi-based chemical sensors.

Inkjet printing has recently been combined with vapor-phase polymerization to achieve patterned PANi [138]. Here, the oxidant (ammonium persulfate) was patterned using inkjet printing. Subsequent exposure of the printed substrate to aniline vapor resulted in the formation of PANi. Films of less than 0.5 μm were formed with line widths down to 80 μm.

A commercial PPy dispersion from Sigma Aldrich (Product Number 482552) was modified by addition of EG (10%) to reduce viscosity to less than 100 mPa s and obtain a surface tension of approximately 35 mN/m, and hence produce a printable formulation [139]. A significant increase in conductivity was observed on exposure of the films to the vapors of simple alcohols, forming the basis of a printable sensing technology.

The inkjet printing of poly-3-alkylthiophenes has been reported [140], although the printing of polythiophenes is made difficult by the volatility of solvents such as chloroform used to dissolve polythiophenes, causing the the jets to clog. This methodology also provides a way to create scaffolds for cell culturing [141].

Printed polythiophenes have also been utilized as chemical sensors; for example, PEDOT:PSS has been used for the detection of organic vapors [142]. The electrical resistance of inkjet-printed films was monitored when exposed to atmospheres containing alcohol. The resistance of thin films (one or two printed layers), increased sharply and nonreversibly as the concentration of alcohol vapor in a carrier gas increased. An intended application for this inkjet-printed device is a handheld instrument to monitor the presence of organic vapors.

Gas sensors have also been inkjet printed using a range of regioregular polythiophenes with different functional groups [143]. Sensors were printed using a custom inkjet deposition system with a Microfab drop-on-demand single-nozzle printhead delivering 50 pl drops through 30 μm nozzles with ±4 μm drop placement accuracy.

Polythiophene biosensors have also been fabricated by thermal inkjet printing of PEDOT:PSS and GOx in sequence onto indium tin oxide (ITO) glass [144] (Figure 5.27). Films of PEDOT and GOx were printed and the resulting biosensors were encapsulated with a cellulose acetate membrane to prevent dissolution of the active layers.

5.6.8
Printing Carbon Nanotubes

Recently, the addition of CNTs to conducting polymer inks to produce printable formulations with improved properties has attracted attention. Transparent and conductive patterns of carboxyl-functionalized SWNTs (SWNT-COOHs) and the composites of those with PEDOT:PSS have been deposited on various substrates by inkjet printing [145]. For low-print repetitions, the PEDOT:PSS/SWNT-COOH composite patterns show enhanced conductance as compared with the corresponding PEDOT:PSS conductors. Patterns with sheet resistivities as low as $\sim 1 \times 10^3\ \Omega\,\mathrm{sq}^{-1}$ were achieved, and while there is a trade-off between transparency and

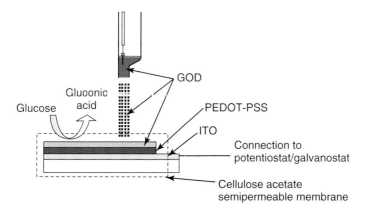

Figure 5.27 Figure 5.27 Diagram of the prototype of the complete GOD inkjet-printed electrode.

conductivity, highly transparent patterns ($\sim$90%) with a reasonably low resistivity of $\sim$1 $\times$ 10^4 Ω sq^{-1} were realized.

Denneulin *et al.* mixed PEDOT:PSS (Baytron P) with polyethylene glycol (PEG) functionalized SWNTs to produce a much more conductive film compared to the original PEDOT:PSS film [146]. Conductive patterns were deposited on polymer films using a Dimatix 2831 inkjet printer. Performances of several CNTs were evaluated (SWNT, MWNT, and functionalized CNTs) and sheet resistances ranged from 10 537 to 225 Ω sq^{-1}. The latter, achieved using CNTs with polyethyleneglycol (PEG-SWNT), appeared to be the best candidates for printed electronics with sheet resistances as low as 225 Ω sq^{-1}; which is one of the lowest resistances obtained by inkjet printing. This study represents an important step for the integration of CNTs in printed electronics applications and offers new opportunities to produce cost-effective electronics.

The synthesis and inkjet processing of a water-dispersible PANi composite ink comprising poly(methoxyanilinesulfonic acid) [23] with MWNT loadings of up to 32% have been described [147, 148]. Dispersions with nanotubes (concentration 10 mg ml^{-1}, temperature 25 $^\circ$C, viscosity 5.5 mPa s, and surface tension 72 mN m^{-1}) were inkjet printed using a Dimatix materials deposition system on PET. Cyclic voltammograms obtained using films printed onto Au-PVDF (polyvinyl difluoride) exhibited three redox couples attributed previously to the following interconversions: PANi/PMAS leucoemeraldine–emeraldine, PMAS emeraldine–pernigraniline, and PANi emeraldine–pernigraniline. The sheet resistance and conductivity of free-standing films at the highest loading fractions were 5 Ω sq^{-1} and 51 S cm^{-1}, respectively (Figure 5.28). The unique combination of conducting electroactive polymers with conducting CNTs has been proved to be an ideal formulation, with all of the demanding characteristics needed for inkjet printing. These materials could be readily deposited onto substrates such as photopaper, PET, Pt-ITO, and Au-PVDF. A sheet resistance of 500 Ω sq^{-1} was attainable for a single printed layer on photopaper.

5.6.9
Printed Electronics as a Basis for Printed Bionics

A range of electronic components based on conducting polymers have been printed. It is obvious that such advances will be captured by the bionics research community. This will enable us to transcend the use of organic conductors as simple organic wires or electrochemical transducers, as these have already been proved for electrical stimulation of living cells. The printed transistor circuit was first studied by Sirringhaus *et al.* [149] in 2000 (Figure 5.29). Since then, a series of all-printed transistors and devices have been fabricated and characterized. Kawase *et al.* [150] fabricated all-polymer

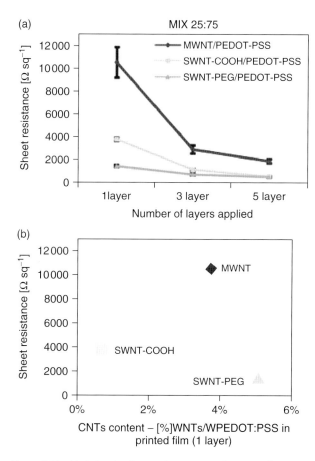

Figure 5.28 Variation in sheet resistance as a function of printed layers (a) and CNT percentage content (b). (Adapted with permission from Small and coworkers [147].)

inkjet-printed PEDOT:PSS thin-film transistors (TFTs) with a desktop inkjet printer. The transfer characteristics of the device exhibited a high on–off ratio of more than 10^5. A study by Mannerbro *et al.* [151] demonstrated that it is possible to use a simple desktop inkjet printer to manufacture organic electrochemical circuitry on rough flexible carrier substrates such as glossy paper. Owing to the ease with which a circuit designer can go from design idea to prototype device using a flexible inkjet printing system, this technique can be anticipated to be used for the production of a series of small devices based on the electrochemical technology discussed [152].

Printing has also been used to produce all-polymer PEDOT:PSS capacitors. A commercial Epson Stylus color 480 SXU printer was used to inkjet print commercial PEDOT:PSS dispersion (Baytron P) as the conductive layers of the capacitor. Poly(biphenyltetracarboxylic

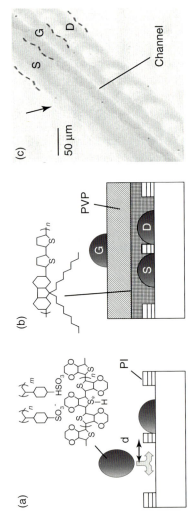

Figure 5.29 Schematic of the high-resolution inkjet printing of source and drain PEDOT electrodes onto a prepatterned substrate (a) and final configuration of the thin-film transistor (b). Optical micrograph of the inkjet-printed thin-film transistor (S, source; D, drain; G, gate) (c). (Adapted with permission from Sirringhaus *et al.* [149].)

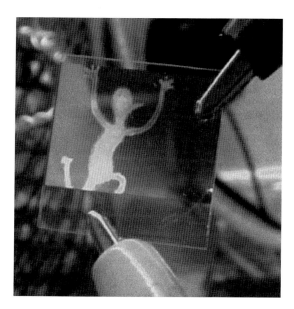

Figure 5.30 Photograph of an organic light-emitting diode (OLED) on a glass substrate where the anode was deposited and patterned via inkjet printing. (Adapted with permission from Yoshioka and Jabbour [156].)

dianhydride-*co*-phenylenediamine) (PBPDA-PD) was chosen as the insulating material [153]. RC filter circuits based on these capacitors have also been successfully fabricated using PEDOT:PSS (Baytron P) and PANi (Sigma) as the conductive layers [154].

Recent developments, however, show that microscale electronics can be made even with readily available inkjet printers. De Rossi and coworkers [155] have used inkjet printing to make three-dimensional structures from soluble conducting polymers. Yoshioka and Jabbour [156] have used a desktop inkjet printer to prepare PEDOT:PSS electrodes as components for organic light-emitting diodes (Figure 5.30). We have prepared PANi nanodispersions that are amenable to printing from a desktop unit [157]. This has been used as a convenient route to produce novel chemical sensing surfaces. Reactive printing, wherein the oxidant is delivered using inkjet printing to produce patterns followed by exposure to monomer in the vapor phase, is an alternative approach to obtain printed patterns [158]. We have used a desktop inkjet printer and an "ink" containing PANi nanoparticles to produce sensors for detection of volatiles [157]. The PANi-based "inks" are attractive in that they print with high resolution and low resistivity [147].

Extrusion printing techniques involving PEDOT:PSS have recently been compared to inkjet printing, with a view to fabricating organic electronics for medical bionics devices [159]. Extruded printed polymer "tracks" were lower in resolution but offered better electrical characteristics compared to inkjet-printed samples. Extrusion printing was also amenable to embedding the PEDOT:PSS patterns into a biopolymer matrix (chitosan and HA) that acts as the material substrate, a feature that again improved the electrical characteristics of the polymer structures. Electrical conductivity of embedded tracks ($17 \, S \, cm^{-1}$) was an order of magnitude higher than for track deposition on the surface of biopolymer film, and three times higher than for tracks on glass. Therefore, extrusion printing offers interesting possibilities in this area, particularly with the ability to fully integrate polymer structures and patterns by embedding them in 3D gel matrices.

These examples serve to illustrate the enormous interest in the development of novel fabrication strategies that enable preparation of low-cost devices based on conducting polymers. The future development of such techniques will undoubtedly lead to devices allowing the integration of organic electronics into a number of conventional host structures. These developments have been hastened by the recent advent of processable conducting polymers. Tremendous opportunities now arise in marrying these processable conducting polymers with the age-old arts of knitting, weaving, and printing.

5.6.10
Dip-Pen Nanolithography

DPN is a nanofabrication technique introduced by Mirkin and coworkers [160] that enables fabrication of devices requiring nanoscale structures and patterns. The technique is low cost, easy to use, and is scalable, enabling fabrication of devices in large quantities. DPN operates by using existing atomic force microscopy (AFM) technology to deposit appropriate formulations onto a surface via a sharp probe tip. The probe tip acts like an "ink pen" by transferring molecules to the surface through a water meniscus, which forms in ambient conditions as the tip nears the surface (Figure 5.31). Dot and line patterns with sizes down to 50 nm can be achieved. "Ink" particles can also be deposited onto the surface via a carrier solution through physioadsorption; however, while this approach is versatile, the pattern resolution is typically lower.

DPN is a nanofabrication tool capable of patterning complex biocomposite materials and devices. By using top-down or bottom-up approaches, the possibility of precisely placing biomolecules and chemical gradients onto existing polymers and fiber materials to control cellular interactions is exciting. Patterning of micro- and nanocircuits based on the deposition of conducting nanomaterials can also be "written" onto medical bionics materials and other devices requiring in-house electronics.

(a)

Ink-coated DPN pen

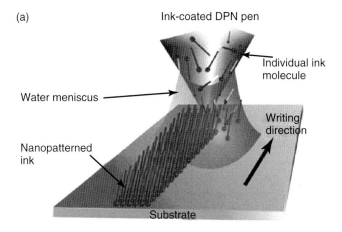

Individual ink
molecule

Water meniscus

Writing
direction

Nanopatterned
ink

Substrate

(b)

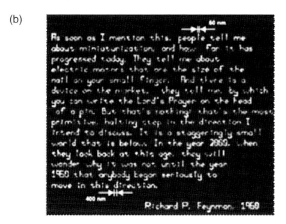

Figure 5.31 (a) Schematic of DPN mechanism for depositing molecules. (b) Excerpt from Richard Feynman's speech "There's plenty of room at the bottom" written with an organic molecular ink using DPN. (Adapted with permission from Piner and coworkers [160].)

DPN represents a new brand of nontraditional fabrication approaches, the development of which is critical for producing complex nanomaterials and nanostructures. It is capable of filling the "nanometre resolution" void in current inkjet/extrusion printing technologies whose patterning is limited to micron-scale resolution. The patterning process can also be scaled up. Arrays with as many as about 11 million pyramid-shaped pens can be brought into contact with substrates to print nanopatterns [161] (Figure 5.32).

DPN has been used to produce a variety of printed structures, including gold nanoparticles [162], silver nanoparticles [163], oligonucleotides, and proteins [164]. Lu and coworkers [165] have used DPN to pattern

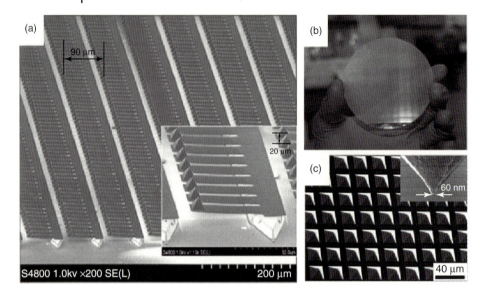

Figure 5.32 (a) SEM image of silicon multi-pen 2D array showing rows of cantilevers. (b) Photograph of polymer-pen array comprising 11 million sharp tips. (c) Magnified image showing individual polymer tips. (Adapted with permission from Huo and coworkers [161].)

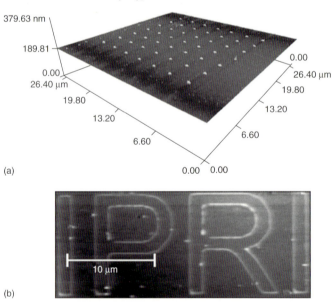

Figure 5.33 (a) Submicron patterning of modified commercial PEDOT:PSS ink using DPN. (b) DPN "writing" of IPRI (Intelligent Polymer Research Institute) using chemical oxidant ink. Line widths ≈100–200 nm.

Table 5.3 Advantages and disadvantages of fabrication techniques.

Fabrication techniques	Advantages	Disadvantages
Drop casting – films	Quick and easy; can produce relatively thick films	Films are not uniform
Dip coating	Reasonably easy	Can be slow and often form tide lines
Spin coating	Fast method for reproducibly producing uniform films	Films are thin and may often contain pinholes and defects
Inkjet printing	Useful for producing complex patterns reproducibly	Volatile solvents can cause clogging of nozzles
Spray drying	Produces powders with consistent particle size distributions; substances can be encapsulated in particles	Limited in the range of materials that can processed
Electrophoretic deposition	Economical method for producing coatings from a wide range of materials	Substrate and deposited material must have reasonable conductivity
Wet spinning – fibers	Good for producing thick fibers	Only produces individual fibers that often need combining
Electrospinning – fibers	Produces fiber mats with fiber diameters of 100 nm to 10 μm	Several parameters need optimizing to avoid beading and defects; requires high molecular weights
Dip-pen nano-lithography	Capable of nanopatterning molecules onto surfaces with dimensions <100 nm	Ink and substrate chemistry can affect deposition and requires optimization; more suited to smooth surfaces

conducting polymer by "writing" lines of a commercial PEDOT:PSS ink and demonstrated their use as nitric oxide gas sensors. Soluble self-doped sulfonated polyaniline (SPANi) and PPy have been patterned with 130 nm line widths onto chemically functionalized substrates to facilitate binding of the ink molecules [166]. DPN has also been used in a manner where the ink is composed of monomer/oxidant constituents and the conducting polymer is written as chemical polymerization occurs *in situ* beneath the probe tip [167].

Electrochemical DPN involves applying a voltage to the probe tip to induce polymerization electrochemically during the printing patterning process. This approach has been used to produce polythiophene materials on insulating and semiconducting substrates [168].

Our laboratories have recently explored the patterning of PEDOT:PSS inks onto a variety of substrates including gold/silicon and flexible substrates such as polyethylene terephthalate and silicone gum. An alternative approach involves printing a chemical oxidant (Fe-*p*TS) with subsequent exposure to the monomer (edot) vapor. Fine control over the patterning of the chemical oxidant is possible, with line widths of 150 nm achieved (Figure 5.33).

5.6.11
Conclusions

The different capabilities of fiber spinning and printing technologies and other techniques such as dip coating, spray drying and layer-by-layer deposition, and DPN (Table 5.3) collectively form an integral and unique set of micro- and nanofabrication tools for manipulating organic conducting materials, as well as soft biopolymer, proteins, and cells. Each of them alone will not necessarily satisfy the requirements for device fabrication, but together they pose a powerful route to assembling complex 3D structures. This is important for medical bionic devices requiring a high level of integration and control over the material properties at the bulk, micro-, and nanodomains.

References

1. Nanko, M. (2009) Definitions and categories of hybrid materials. *AZojomo.*, **6**, 2240–2246.
2. Ramakrishna, S. (2004) Series on biomaterials and bioengineering, in *An Introduction to Biocomposites* (ed. S. Ramakrishna), Imperial College Press, London, 1–17.
3. Liu, Y., Liu, X., Chen, J., Gilmore, K.J., and Wallace, G.G. (2008) 3D Bio-nanofibrous PPy/SIBS mats as platforms for cell culturing. *Chem. Commun.*, 3729–3731.
4. Li, M., Mondrinos, M.J., Gandhi, M.R., Ko, F.K., Weiss, A.S., and Lelkes, P.I. (2005) Electrospun protein fibers as matrices for tissue engineering. *Biomaterials*, **26**, 5999–6008.
5. Angelopoulos, M., Asturias, G.E., Ermer, S.P., Ray, A., Scherr, E.M., Macdiarmid, A.G., Akhtar, M., Kiss, Z., and Epstein, A.J. (1988) Polyaniline: solutions, films and oxidation state. *Mol. Cryst. Liq. Cryst.*, **160**, 151–163.
6. Cao, Y., Smith, P., and Heeger, A.J. (1990) Conjugated bionic materials, in *Conjugated Polymeric Materials: Opportunities in Electronics, Optoelectronics, and Molecular Electronics* (eds J.L. Bredas and R.R. Chance) Kluwer Academic Publishers, Doredecht.
7. Andreatta, A., Cao, Y., Chiang, J.C., Heeger, A.J., and Smith, P. (1988) Electrically-conductive fibers of polyaniline spun from solutions in concentrated sulfuric acid. *Synth. Met.*, **26**, 383–389.
8. Cao, Y., Smith, P., and Heeger, A.J. (1992) Counter-ion induced processibility of conducting polyaniline and of conducting polyblends of polyaniline in bulk polymers. *Synth. Met.*, **48**, 91–97.

9. Heeger, A.J. (1993) Polyaniline with surfactant counterions: conducting polymer materials which are processible in the conducting form. *Synth. Met.*, **57**, 3471–3482.

10. Pron, A., Nicolau, Y., Genoud, F., and Nechtschein, M. (1997) Flexible, highly transparent, and conductive polyaniline-cellulose acetate composite films. *J. Appl. Polym. Sci.*, **63**, 971–977.

11. Gazotti, W.A., Faez, R., and De Paoli, M.A. (1999) Thermal and mechanical behaviour of a conductive elastomeric blend based on a soluble polyaniline derivative. *Eur. Polym. J.*, **35**, 35–40.

12. Su, S.J. and Kuramoto, N. (2000) Synthesis of processable polyaniline complexed with anionic surfactant and its conducting blends in aqueous and organic system. *Synth. Met.*, **108**, 121–126.

13. Rühe, J., Ezquerra, T., and Wegner, G. (1989) Conducting polymers from 3-alkylpyrroles. *Makromol. Chem. Rapid Commun.*, **10**, 103–108.

14. Nalwa, H.S. (1991) Chemical synthesis of processible electrically conducting poly(3-dodecylthiophene). *Angew. Makromol. Chem.*, **188**, 105–111.

15. Ashraf, S.A., Chen, F., Too, C.O., and Wallace, G.G. (1996) Bulk electropolymerization of alkylpyrroles. *Polymer*, **37**, 2811–2819.

16. Faid, K., Cloutier, R., and Leclerc, M. (1993) Design of new electroactive polybithiophene derivatives. *Synth. Met.*, **55**, 1272–1277.

17. Patil, A.O., Ikenoue, Y., Basescu, N., Colaneri, N., Chen, J., Wudl, F., and Heeger, A.J. (1987) Self-doped conducting polymers. *Synth. Met.*, **20**, 151–159.

18. Ikenoue, Y., Outani, N., Patil, A.O., Wudl, F., and Heeger, A.J. (1989) Electrochemical studies of self-doped conducting polymers: verification of the cation-hopping doping mechanism. *Synth. Met.*, **30**, 305–319.

19. Ikenoue, Y., Wudl, F., and Heeger, A.J. (1991) A novel substituted poly(isothianaphthene). *Synth. Met.*, **40**, 1–12.

20. Yue, J., Wang, Z.H., Cromack, K.R., Epstein, A.J., and MacDiarmid, A.G. (1991) Effect of sulfonic acid group on polyaniline backbone. *J. Am. Chem. Soc.*, **113**, 2665–2671.

21. Shimizu, S., Saitoh, T., Uzawa, M., Yuasa, M., Yano, K., Maruyama, T., and Watanabe, K. (1997) Synthesis and applications of sulfonated polyaniline. *Synth. Met.*, **85**, 1337–1338.

22. Guo, R., Barisci, J.N., Innis, P.C., Too, C.O., Wallace, G.G., and Zhou, D. (2000) Electrohydrodynamic polymerization of 2-methoxyaniline-5-sulfonic acid. *Synth. Met.*, **114**, 267–272.

23. Zhou, D., Innis, P.C., Wallace, G.G., Shimizu, S., and Maeda, S.I. (2000) Electrosynthesis and characterisation of poly(2-methoxyaniline-5-sulfonic acid)-effect of pH control. *Synth. Met.*, **114**, 287–293.

24. Masdarolomoor, F., Innis, P.C., Ashraf, S., and Wallace, G.G. (2005) Purification and characterisation of poly(2-methoxyaniline-5-sulfonic acid). *Synth. Met.*, **153**, 181–184.

25. Aldissi, M. and Armes, S.P. (1991) Colloidal dispersions of conducting polymers. *Prog. Org. Coat.*, **19**, 21–58.

26. Eisazadeh, H.S., Spinks, G., and Wallace, G.G. (1993) Conductive electroactive paint containing polypyrrole colloids. *Mater. Forum*, **17**, 241–245.

27. Bose, C.S.C., Basak, S., and Rajeshwar, K. (1992) Preparation, voltammetric characterization, and use of a composite containing chemically synthesized polypyrrole and a carrier polymer. *J. Electrochem. Soc.*, **139**, L75–L76.

28. Barisci, J.N., Hodgson, A.J., Liu, L., Wallace, G.G., and Harper, G. (1999) Electrochemical production

of protein-containing polypyrrole colloids. *React. Funct. Polym.*, **39**, 269–275.

29. Eisazadeh, H., Gilmore, K.J., Hodgson, A.J., Spinks, G., and Wallace, G.G. (1995) Electrochemical production of conducting polymer colloids. *Colloids Surf., A*, **103**, 281–288.

30. Gangopadhyay, R. and De, A. (2000) Conducting polymer nanocomposites:? A brief overview. *Chem. Mater.*, **12**, 2064.

31. Wallace, G.G. and Innis, P.C. (2002) Inherently conducting polymer nanostructures. *J. Nanosci. Nanotechnol.*, **2**, 441–451.

32. Armes, S.P., Gottesfeld, S., Beery, J.G., Garzon, F., and Agnew, S.F. (1991) Conducting polymer-colloidal silica composites. *Polymer*, **32**, 2325–2330.

33. Vasilyeva, S.V., Vorotyntsev, M.A., Bezverkhyy, I., Lesniewska, E., Heintz, O., and Chassagnon, R. (2008) Synthesis and characterization of palladium nanoparticle/polypyrrole composites. *J. Phys. Chem. C*, **112**, 19878–19885.

34. Salsamendi, M., Marcilla, R., Döbbelin, M., Mecerreyes, D., Pozo-Gonzalo, C., Pomposo, J.A., and Pacios, R. (2008) Simultaneous synthesis of gold nanoparticles and conducting poly(3,4-ethylenedioxythiophene) towards optoelectronic nanocomposites. *Phys. Status Solidi A*, **205**, 1451–1454.

35. Yin, K. and Zhu, Z. (2010) "One-pot" synthesis, characterization, and NH_3 sensing of Pd/PEDOT:PSS nanocomposite. *Synth. Met.*, **160**, 1115–1118.

36. Aboutanos, V., Kane-Maguire, L.A.P., and Wallace, G.G. (2000) Electrosynthesis of polyurethane-based core-shell PAn(+)-HCSA colloids. *Synth. Met.*, **114**, 313–320.

37. Tian, S., Liu, J., Zhu, T., and Knoll, W. (2004) Polyaniline/gold nanoparticle multilayer films:? Assembly, properties, and biological

applications. *Chem. Mater.*, **16**, 4103–4108.

38. Paul, R.K. and Pillai, C.K.S. (2001) Melt/solution processable polyaniline with functionalized phosphate ester dopants and its thermoplastic blends. *J. Appl. Polym. Sci.*, **80**, 1354–1367.

39. Paul, R.K. and Pillai, C.K.S. (2001) Thermal properties of processable polyaniline with novel sulfonic acid dopants. *Polym. Int.*, **50**, 381–386.

40. Paul, R.K., Vijayanathan, V., and Pillai, C.K.S. (1999) Melt/solution processable conducting polyaniline: doping studies with a novel phosphoric acid ester. *Synth. Met.*, **104**, 189–195.

41. Qian, D., Dickey, E.C., Andrews, R., Rantell, T., Jacques, D., Rao, A.M., Derbyshire, F., Chen, Y., Chen, J., and Haddon, R.C. (2000) Load transfer and deformation mechanisms in carbon nanotube-polystyrene composites. Nanotube composite carbon fibers. *Appl. Phys. Lett.*, **76**, 2868–2870.

42. Liu, J., Rasheed, A., Minus, M.L., and Kumar, S. (2009) Processing and properties of carbon nanotube/poly(methyl methacrylate) composite films. *J. Appl. Polym. Sci.*, **112**, 142–156.

43. Gilmore, K.J., Moulton, S.E., and Wallace, G.G. (2007) Incorporation of carbon nanotubes into the biomedical polymer poly(styrene-[beta]-isobutylene-[beta]-styrene). *Carbon*, **45**, 402–410.

44. Kearns, J.C. and Shambaugh, R.L. (2002) Polypropylene fibers reinforced with carbon nanotubes. *J. Appl. Polym. Sci.*, **86**, 2079–2084.

45. Safadi, B., Andrews, R., and Grulke, E.A. (2002) Multiwalled carbon nanotube polymer composites: synthesis and characterization of thin films. *J. Appl. Polym. Sci.*, **84**, 2660–2669.

46. Shaffer, M.S.P. and Windle, A.H. (1999) Analogies between polymer solutions and carbon nanotube dispersions. *Macromolecules*, **32**, 6864–6866.

47. Shaffer, M.S.P. and Windle, A.H. (1999) Fabrication and characterization of carbon nanotube/poly(vinyl alcohol) composites. *Adv. Mater.*, **11**, 937–941.

48. Coleman, J.N., O'Brien, D.F., Dalton, A.B., McCarthy, B., Lahr, B., Barklie, R.C., and Blau, W.J. (2000) Electron paramagnetic resonance as a quantitative tool for the study of multiwalled carbon nanotubes. *J. Chem. Phys.*, **113**, 9788–9793.

49. Gong, X., Liu, J., Baskaran, S., Voise, R.D., and Young, J.S. (2000) Surfactant-assisted processing of carbon nanotube/polymer composites. *Chem. Mater.*, **12**, 1049–1052.

50. Bryning, M.B., Milkie, D.E., Islam, M.F., Kikkawa, J.M., Yodh, A.G., Zhang, J., Alsayed, A., Lin, K.H., Sanyal, S., Zhang, F. *et al.* (2005) Thermal conductivity and interfacial resistance in single-wall carbon nanotube epoxy composites. *Appl. Phys. Lett.*, **87**, 161909–161903.

51. Sundararajan, P.R., Singh, S., and Moniruzzaman, M. (2004) Surfactant-induced crystallization of polycarbonate. *Macromolecules*, **37**, 10208–10211.

52. Barrau, S., Demont, P., Perez, E., Peigney, A., Laurent, C., and Lacabanne, C. (2003) Effect of palmitic acid on the electrical conductivity of carbon nanotubes-epoxy resin composites. *Macromolecules*, **36**, 9678–9680.

53. Barraza, H.J., Pompeo, F., O'Rea, E.A., and Resasco, D.E. (2002) SWNT-filled thermoplastic and elastomeric composites prepared by miniemulsion polymerization. *Nano Lett.*, **2**, 797–802.

54. Raravikar, N.R., Schadler, L.S., Vijayaraghavan, A., Zhao, Y., Wei, B., and Ajayan, P.M. (2005) Synthesis and characterization of thickness-aligned carbon nanotube-polymer composite films. *Chem. Mater.*, **17**, 974–983.

55. Ramasubramaniam, R., Chen, J., and Liu, H. (2003) Homogeneous carbon nanotube/polymer composites for electrical applications. *Appl. Phys. Lett.*, **83**, 2928–2930.

56. Mitchell, C.A., Bahr, J.L., Arepalli, S., Tour, J.M., and Krishnamoorti, R. (2002) Dispersion of functionalized carbon nanotubes in polystyrene. *Macromolecules*, **35**, 8825–8830.

57. Hill, D.E., Lin, Y., Rao, A.M., Allard, L.F., and Sun, Y.P. (2002) Functionalization of carbon nanotubes with polystyrene. *Macromolecules*, **35**, 9466–9471.

58. Zhu, J., Peng, H., Rodriguez-Macias, F., Margrave, J.L., Khabashesku, V.N., Imam, A.M., Lozano, K., and Barrera, E.V. (2004) Reinforcing epoxy polymer composites through covalent integration of functionalized nanotubes. *Adv. Funct. Mater.*, **14**, 643–648.

59. Yu, R., Chen, L., Liu, Q., Lin, J., Tan, K.L., Ng, S.C., Chan, H.S.O., Xu, G.Q., and Hor, T.S.A. (1998) Platinum deposition on carbon nanotubes via chemical modification. *Chem. Mater.*, **10**, 718–722.

60. Jiang, K., Eitan, A., Schadler, L.S., Ajayan, P.M., Siegel, R.W., Grobert, N., Mayne, M., Reyes-Reyes, M., Terrones, H., and Terrones, M. (2003) Selective attachment of gold nanoparticles to nitrogen-doped carbon nanotubes. *Nano Lett.*, **3**, 275–277.

61. Hu, X., Wang, T., Qu, X., and Dong, S. (2005) In situ synthesis and characterization of multiwalled carbon nanotube/Au nanoparticle composite materials. *J. Phys. Chem. B*, **110**, 853–857.

62. Yang, M., Jiang, J., Yang, Y., Chen, X., Shen, G., and Yu, R. (2006) Carbon nanotube/cobalt hexacyanoferrate nanoparticle-biopolymer system for the fabrication of biosensors. *Biosens. Bioelectron.*, **21**, 1791–1797.

63. Hrapovic, S., Liu, Y., Male, K.B., and Luong, J.H.T. (2003) Electrochemical biosensing platforms using platinum nanoparticles and

carbon nanotubes. *Anal. Chem.*, **76**, 1083–1088.

64. Georgakilas, V., Gournis, D., Tzitzios, V., Pasquato, L., Guldi, D.M., and Prato, M. (2007) Decorating carbon nanotubes with metal or semiconductor nanoparticles. *J. Mater. Chem.*, **17**, 2679–2694.

65. Liu, T., Phang, I.Y., Shen, L., Chow, S.Y., and Zhang, W.D. (2004) Morphology and mechanical properties of multiwalled carbon nanotubes reinforced nylon-6 composites. *Macromolecules*, **37**, 7214–7222.

66. Pötschke, P., Bhattacharyya, A.R., and Janke, A. (2004) Carbon nanotube-filled polycarbonate composites produced by melt mixing and their use in blends with polyethylene. *Carbon*, **42**, 965–969.

67. Pötschke, P., Dudkin, S.M., and Alig, I. (2003) Dielectric spectroscopy on melt processed polycarbonate–multiwalled carbon nanotube composites. *Polymer*, **44**, 5023–5030.

68. Zhang, Q., Rastogi, S., Chen, D., Lippits, D., and Lemstra, P.J. (2006) Low percolation threshold in single-walled carbon nanotube/high density polyethylene composites prepared by melt processing technique. *Carbon*, **44**, 778–785.

69. Haggenmueller, R., Gommans, H.H., Rinzler, A.G., Fischer, J.E., and Winey, K.I. (2000) Aligned single-wall carbon nanotubes in composites by melt processing methods. *Chem. Phys. Lett.*, **330**, 219–225.

70. Andrews, R., Jacques, D., Minot, M., and Rantell, T. (2002) Fabrication of carbon multiwall nanotube/polymer composites by Shear mixing. *Macromol. Mater. Eng.*, **287**, 395–403.

71. Coleman, J.N., Khan, U., Blau, W.J., and Gun'ko, Y.K. (2006) Small but strong: a review of the mechanical properties of carbon nanotube-polymer composites. *Carbon*, **44**, 1624–1652.

72. Stankovich, S., Dikin, D.A., Dommett, G.H.B., Kohlhaas, K.M., Zimney, E.J., Stach, E.A., Piner, R.D., Nguyen, S.T., and Ruoff, R.S. (2006) Graphene-based composite materials. *Nature*, **442**, 282–286.

73. Eda, G. and Chhowalla, M. (2009) Graphene-based composite thin films for electronics. *Nano Lett.*, **9**, 814–818.

74. Ramanathan, T., Abdala, A.A., Stankovich, S., Dikin, D.A., Herrera-Alonso, M., Piner, R.D., Adamson, D.H., Schniepp, H.C., Chen, X., Ruoff, R.S. *et al.* (2008) Functionalized graphene sheets for polymer nanocomposites. *Nat. Nanotechnol.*, **3**, 327–331.

75. Kim, H. and Macosko, C.W. (2009) Processing-property relationships of polycarbonate/graphene composites. *Polymer*, **50**, 3797–3809.

76. Weng, W., Chen, G., Wu, D., Chen, X., Lu, J., and Wang, P. (2004) Fabrication and characterization of nylon 6/foliated graphite electrically conducting nanocomposite. *J. Polym. Sci. Part B Polym. Phys.*, **42**, 2844–2856.

77. Kim, H., Kobayashi, S., AbdurRahim, M.A., Zhang, M.J., Khusainova, A., Hillmyer, M.A., Abdala, A.A., and Macosko, C.W. (2011) Graphene/polyethylene nanocomposites: effect of polyethylene functionalization and blending methods. *Polymer*, **52**, 1837–1846.

78. Lee, H.B., Raghu, A.V., Yoon, K.S., and Jeong, H.M. (2010) Preparation and characterization of poly(ethylene oxide)/graphene nanocomposites from an aqueous medium. *J. Macromol. Sci. Part B Phys.*, **49**, 802–809.

79. Tkalya, E., Ghislandi, M., Alekseev, A., Koning, C., and Loos, J. (2010) Latex-based concept for the preparation of graphene-based polymer nanocomposites. *J. Mater. Chem.*, **20**, 3035–3039.

80. Jang, B. and Zhamu, A. (2008) Processing of nanographene platelets

(NGPs) and NGP nanocomposites: a review. *J. Mater. Sci.*, **43**, 5092–5101.

81. Keohan, F., Wei, X.F., Wongsarnpigoon, A., Lazaro, E., Darga, J.E., and Grill, W.M. (2007) Fabrication and evaluation of conductive elastomer electrodes for neural stimulation. *J. Biomater. Sci. Polym. Ed.*, **18**, 1057–1073.

82. Zhang, Z., Roy, R., Dugré, F.J., Tessier, D., and Dao, L.H. (2001) In vitro biocompatibility study of electrically conductive polypyrrole-coated polyester fabrics. *J. Biomed. Mater. Res.*, **57**, 63–71.

83. Cucchi, I., Boschi, A., Arosio, C., Bertini, F., Freddi, G., and Catellani, M. (2009) Bio-based conductive composites: preparation and properties of polypyrrole (PPy)-coated silk fabrics. *Synth. Met.*, **159**, 246–253.

84. Koyama, S., Haniu, H., Osaka, K., Koyama, H., Kuroiwa, N., Endo, M., Kim, Y.A., and Hayashi, T. (2006) Medical application of carbon-nanotube-filled nanocomposites: the microcatheter. *Small*, **2**, 1406–1411.

85. Razal, J.M., Kita, M., Quigley, A.F., Kennedy, E., Moulton, S.E., Kapsa, R.M.I., Clark, G.M., and Wallace, G.G. (2009) Wet-spun biodegradable fibers on conducting platforms: novel architectures for muscle regeneration. *Adv. Funct. Mater.*, **19**, 3381–3388.

86. Knill, C.J., Kennedy, J.F., Mistry, J., Miraftab, M., Smart, G., Groocock, M.R., and Williams, H.J. (2004) Alginate fibres modified with unhydrolysed and hydrolysed chitosans for wound dressings. *Carbohydr. Polym.*, **55**, 65–76.

87. Matsuda, A., Ikoma, T., Kobayashi, H., and Tanaka, J. (2004) Preparation and mechanical property of core-shell type chitosan/calcium phosphate composite fiber. *Mater. Sci. Eng., C*, **24**, 723–728.

88. Vigolo, B., Pénicaud, A., Coulon, C., Sauder, C., Pailler, R., Journet, C., Bernier, P., and Poulin, P.

(2000) Macroscopic fibers and ribbons of oriented carbon nanotubes. *Science*, **290**, 1331–1334.

89. Lynam, C., Moulton, S.E., and Wallace, G.G. (2007) Carbon-nanotube biofibers. *Adv. Mater.*, **19**, 1244–1248.

90. Laughlin, P.J. and Monkman, A.P. (1997) Mechanical properties of oriented emeraldine base polyaniline. *Synth. Met.*, **84**, 765–766.

91. Adams, P.N., Bowman, D., Brown, L., Yang, D., and Mattes, B.R. (2001) Molecular weight dependence of the physical properties of protonated polyaniline films and fibers. *Proc. SPIE Int. Soc. Opt. Eng.*, **4329**, 475.

92. Wang, H.L., Romero, R.J., Mattes, B.R., Zhu, Y., and Winokur, M.J. (2000) Effect of processing conditions on the properties of high molecular weight conductive polyaniline fiber. *J. Polym. Sci. Part B Polym. Phys.*, **38**, 194–204.

93. Pomfret, S.J., Adams, P.N., Comfort, N.P., and Monkman, A.P. (2000) Electrical and mechanical properties of polyaniline fibres produced by a one-step wet spinning process. *Polymer*, **41**, 2265–2269.

94. Mottaghitalab, V., Spinks, G.M., and Wallace, G.G. (2006) The development and characterisation of polyaniline--single walled carbon nanotube composite fibres using 2-acrylamido-2 methyl-1-propane sulfonic acid (AMPSA) through one step wet spinning process. *Polymer*, **47**, 4996–5002.

95. Mottaghitalab, V., Xi, B., Spinks, G.M., and Wallace, G.G. (2006) Polyaniline fibres containing single walled carbon nanotubes: enhanced performance artificial muscles. *Synth. Met.*, **156**, 796–803.

96. Takahashi, T., Ishihara, M., and Okuzaki, H. (2005) Poly(3,4-ethylenedioxythiophene)/poly(4-styrenesulfonate) microfibers. *Synth. Met.*, **152**, 73–76.

97. Okuzaki, H., Harashina, Y., and Yan, H. (2009) Highly conductive PEDOT/PSS microfibers fabricated

by wet-spinning and dip-treatment in ethylene glycol. *Eur. Polym. J.*, **45**, 256–261.

98. Foroughi, J., Spinks, G.M., Wallace, G.G., and Whitten, P.G. (2008) Production of polypyrrole fibres by wet spinning. *Synth. Met.*, **158**, 104–107.

99. Foroughi, J., Ghorbani, S.R., Peleckis, G., Spinks, G.M., Wallace, G.G., Wang, X.L., and Dou, S.X. (2010) The mechanical and the electrical properties of conducting polypyrrole fibers. *J. Appl. Phys.*, **107**, 103712–103714.

100. Foroughi, J., Spinks, G.M., and Wallace, G.G. (2011) A reactive wet spinning approach to polypyrrole fibres. *J. Mater. Chem.*, **21**, 6421–6426.

101. Steinmetz, J., Glerup, M., Paillet, M., Bernier, P., and Holzinger, M. (2005) Production of pure nanotube fibers using a modified wet-spinning method. *Carbon*, **43**, 2397–2400.

102. Kumar, S., Dang, T.D., Arnold, F.E., Bhattacharyya, A.R., Min, B.G., Zhang, X., Vaia, R.A., Park, C., Adams, W.W., Hauge, R.H. *et al.* (2002) Synthesis, structure, and properties of PBO/SWNT composites. *Macromolecules*, **35**, 9039–9043.

103. Razal, J.M., Gilmore, K.J., and Wallace, G.G. (2008) Carbon nanotube biofiber formation in a polymer-free coagulation bath. *Adv. Funct. Mater.*, **18**, 61–66.

104. Spinks, G.M., Shin, S.R., Wallace, G.G., Whitten, P.G., Kim, I.Y., Kim, S.I., and Kim, S.J. (2007) A novel "dual mode" actuation in chitosan/polyaniline/carbon nanotube fibers. *Sens. Actuators, B*, **121**, 616–621.

105. Granero, A.J., Razal, J.M., Wallace, G.G., and in het Panhuis, M. (2010) Conducting gel-fibres based on carrageenan, chitosan and carbon nanotubes. *J. Mater. Chem.*, **20**, 7953–7956.

106. Granero, A.J., Razal, J.M., Wallace, G.G., and in het Panhuis, M. (2008) Spinning carbon nanotube-gel fibers using poly-electrolyte complexation. *Adv. Funct. Mater.*, **18**, 3759–3764.

107. Quigley, A.F., Razal, J.M., Thompson, B.C., Moulton, S.E., Kita, M., Kennedy, E.L., Clark, G.M., Wallace, G.G., and Kapsa, R.M.I. (2009) A conducting-polymer platform with biodegradable fibers for stimulation and guidance of axonal growth. *Adv. Mater.*, **21**, 4393–4397.

108. Reneker, D.H. and Chun, I. (1996) Nanometre diameter fibres of polymer, produced by electrospinning. *Nanotechnology*, **7**, 216.

109. Theron, A. *et al.* (2001) Electrostatic field-assisted alignment of electrospun nanofibres. *Nanotechnology*, **12**, 384.

110. MacDiarmid, A.G., Jones, W.E. Jr., Norris, I.D., Gao, J., Johnson, A.T. Jr., Pinto, N.J., Hone, J., Han, B., Ko, F.K., Okuzaki, H. *et al.* (2001) Electrostatically-generated nanofibers of electronic polymers. *Synth. Met.*, **119**, 27–30.

111. Huber, A., Pickett, A., and Shakesheff, K.M. (2007) Reconstruction of spatially orientated myotubes in vitro using electrospun, parallel microfibre arrays. *Eur. Cells Mater.*, **14**, 56–63.

112. Kang, T.S., Lee, S.W., Joo, J., and Lee, J.Y. (2005) Electrically conducting polypyrrole fibers spun by electrospinning. *Synth. Met.*, **153**, 61–64.

113. Nair, S., Natarajan, S., and Kim, S.H. (2005) Fabrication of electrically conducting polypyrrole-poly(ethylene oxide) composite nanofibers. *Macromol. Rapid Commun.*, **26**, 1599–1603.

114. Liu, X., Chen, J., Gilmore, K.J., Higgins, M.J., Liu, Y., and Wallace, G.G. (2010) Guidance of neurite outgrowth on aligned electrospun polypyrrole/poly(styrene-β-isobutylene-β-styrene) fiber platforms. *J. Biomed. Mater. Res. Part A*, **94A**, 1004–1011.

115. Chronakis, I.S., Grapenson, S., and Jakob, A. (2006) Conductive polypyrrole nanofibers via electrospinning: electrical and morphological properties. *Polymer*, **47**, 1597–1603.

116. Babel, A., Li, D., Xia, Y., and Jenekhe, S.A. (2005) Electrospun nanofibers of blends of conjugated polymers:? Morphology, optical properties, and field-effect transistors. *Macromolecules*, **38**, 4705–4711.

117. Bianco, A., Bertarelli, C., Frisk, S., Rabolt, J.F., Gallazzi, M.C., and Zerbi, G. (2007) Electrospun polyalkylthiophene/ polyethyleneoxide fibers: optical characterization. *Synth. Met.*, **157**, 276–281.

118. Liu, H., Reccius, C.H., and Craighead, H.G. (2005) Single electrospun regioregular poly(3-hexylthiophene) nanofiber field-effect transistor. *Appl. Phys. Lett.*, **87**, 253106–253103.

119. Li, D., Babel, A., Jenekhe, S.A., and Xia, Y. (2004) Nanofibers of conjugated polymers prepared by electrospinning with a two-capillary spinneret. *Adv. Mater.*, **16**, 2062–2066.

120. González, R. and Pinto, N.J. (2005) Electrospun poly(3-hexylthiophene-2,5-diyl) fiber field effect transistor. *Synth. Met.*, **151**, 275–278.

121. Laforgue, A. and Robitaille, L. (2008) Fabrication of poly-3-hexylthiophene/polyethylene oxide nanofibers using electrospinning. *Synth. Met.*, **158**, 577–584.

122. Nguyen, H.D., Ko, J.M., Kim, H.J., Kim, S.K., Cho, S.H., Nam, J.D., and Lee, J.Y. (2008) Electrochemical properties of poly(3,4-ethylenedioxythiophene) Nanofiber non-woven web formed by electrospinning. *J. Nanosci. Nanotechnol.*, **8**, 4718–4721.

123. Laforgue, A. and Robitaille, L. (2010) Production of conductive PEDOT nanofibers by the combination of electrospinning and vapor-phase polymerization. *Macromolecules*, **43**, 4194–4200.

124. Abidian, M.R., Kim, D.H., and Martin, D.C. (2006) Conducting-polymer nanotubes for controlled drug release. *Adv. Mater.*, **18**, 405–409.

125. Panapoy, M., Saengsil, M., and Ksapabutr, B. (2008) Electrical conductivity of Poly(3,4-Ethylenedioxythiophene)-Poly(Styrenesulfonate) coatings on polyacrylonitrile nanofibers for sensor applications. *Adv. Mater. Res.*, **55–57**, 257.

126. Breukers, R.D., Gilmore, K.J., Kita, M., Wagner, K.K., Higgins, M.J., Moulton, S.E., Clark, G.M., Officer, D.L., Kapsa, R.M.I., and Wallace, G.G. (2010) Creating conductive structures for cell growth: growth and alignment of myogenic cell types on polythiophenes. *J. Biomed. Mater. Res. Part A*, **95A**, 256–268.

127. Huang, Z.M., Zhang, Y.Z., Kotaki, M., and Ramakrishna, S. (2003) A review on polymer nanofibers by electrospinning and their applications in nanocomposites. *Compos. Sci. Technol.*, **63**, 2223–2253.

128. Xie, X.L., Mai, Y.W., and Zhou, X.P. (2005) Dispersion and alignment of carbon nanotubes in polymer matrix: a review. *Mater. Sci. Eng. R*, **49**, 89–112.

129. Yeo, L.Y. and Friend, J.R. (2006) Electrospinning carbon nanotube polymer composite nanofibers. *J. Exp. Nanosci.*, **1**, 177–209.

130. Liu, Y., Gilmore, K.J., Chen, J., Misoska, V., and Wallace, G.G. (2003) Bio-nanowebs based on Poly(styrene-β-isobutylene-β-styrene) (SIBS) containing single wall carbon nanotubes. *Chem. Mater.*, **19**, 2721–2723.

131. Liu, Y., Chen, J., Anh, N.T., Too, C.O., Misoska, V., and Wallace, G.G. (2008) *Nanofiber Mats from DNA, SWNTs, and Poly(ethylene oxide) and their Application in Glucose Biosensors*, vol. **155**, ETATS-UNIS, Electrochemical Society, Pennington, NJ.

132. Pettersson, L.A.A., Ghosh, S., and Inganäs, O. (2002) Optical anisotropy in thin films of poly(3,4-ethylenedioxythiophene)-poly(4-styrenesulfonate). *Org. Electron.*, **3**, 143–148.

133. de Gans, B.J., Duineveld, P.C., and Schubert, U.S. (2004) Inkjet printing of polymers: state of the art and future developments. *Adv. Mater.*, **16**, 203–213.

134. Ngamna, O., Morrin, A., Killard, A.J., Moulton, S.E., Smyth, M.R., and Wallace, G.G. (2007) Inkjet printable polyaniline nanoformulations. *Langmuir*, **23**, 8569–8574.

135. Morrin, A., Wilbeer, F., Ngamna, O., Moulton, S.E., Killard, A.J., Wallace, G.G., and Smyth, M.R. (2005) Novel biosensor fabrication methodology based on processable conducting polyaniline nanoparticles. *Electrochem. Commun.*, **7**, 317–322.

136. Morrin, A., Ngamna, O., O'Malley, E., Kent, N., Moulton, S.E., Wallace, G.G., Smyth, M.R., and Killard, A.J. (2008) The fabrication and characterization of inkjet-printed polyaniline nanoparticle films. *Electrochim. Acta*, **53**, 5092–5099.

137. Jang, J., Ha, J., and Cho, J. (2007) Fabrication of water-dispersible polyaniline-poly(4-styrenesulfonate) nanoparticles for inkjet-printed chemical-sensor applications. *Adv. Mater.*, **19**, 1772–1775.

138. Cho, J., Shin, K.H., and Jang, J. (2010) Polyaniline micropattern onto flexible substrate by vapor deposition polymerization-mediated inkjet printing. *Thin Solid Films*, **518**, 5066–5070.

139. Mabrook, M.F., Pearson, C., and Petty, M.C. (2006) Inkjet-printed polypyrrole thin films for vapour sensing. *Sens. Actuators, B*, **115**, 547–551.

140. Chang, J.B., Liu, V., Subramanian, V., Sivula, K., Luscombe, C., Murphy, A., Liu, J., and Frechet, J.M.J. (2006) Printable polythiophene gas sensor array for low-cost electronic noses. *J. Appl. Phys.*, **100**, 014506–014507.

141. Campbell, P.G. and Weiss, L.E. (2007) Tissue engineering with the aid of inkjet printers. *Expert Opin. Biol. Ther.*, **7**, 1123–1127.

142. Mabrook, M.F., Pearson, C., Petty, M.C., Ahn, J.H., Wang, C., Bryce, M.R., Christie, P., and Roberts, G.G. (2005) An inkjet-printed chemical fuse. *Appl. Phys. Lett.*, **86**, 013507–013510.

143. Li, B., Santhanam, S., Schultz, L., Jeffries-El, M., Iovu, M.C., Sauvé, G., Cooper, J., Zhang, R., Revelli, J.C., Kusne, A.G. *et al.* (2007) Inkjet printed chemical sensor array based on polythiophene conductive polymers. *Sens. Actuators, B*, **123**, 651–660.

144. Setti, L., Fraleoni-Morgera, A., Ballarin, B., Filippini, A., Frascaro, D., and Piana, C. (2005) An amperometric glucose biosensor prototype fabricated by thermal inkjet printing. *Biosens. Bioelectron.*, **20**, 2019–2026.

145. Mustonen, T., Kordás, K., Saukko, S., Tóth, G., Penttilä, J.S., Helistö, P., Seppä, H., and Jantunen, H. (2007) Inkjet printing of transparent and conductive patterns of single-walled carbon nanotubes and PEDOT-PSS composites. *Phys. Status Solidi B*, **244**, 4336–4340.

146. Aurore, D. *et al.* (2009) The influence of carbon nanotubes in inkjet printing of conductive polymer suspensions. *Nanotechnology*, **20**, 385701.

147. Small, W.R., Masdarolomoor, F., Wallace, G.G., and in het Panhuis, M. (2007) Inkjet deposition and characterization of transparent conducting electroactive polyaniline composite films with a high carbon nanotube loading fraction. *J. Mater. Chem.*, **17**, 4359–4361.

148. Small, W.R. and in het Panhuis, M. (2007) Inkjet printing of transparent, electrically conducting single-walled carbon-nanotube composites. *Small*, **3**, 1500–1503.

149. Sirringhaus, H., Kawase, T., Friend, R.H., Shimoda, T., Inbasekaran, M., Wu, W., and Woo, E.P. (2000) High-resolution inkjet printing of all-polymer transistor circuits. *Science*, **290**, 2123–2126.

150. Kawase, T., Shimoda, T., Newsome, C., Sirringhaus, H., and Friend, R.H. (2003) Inkjet printing of polymer thin film transistors. *Thin Solid Films*, **438–439**, 279–287.

151. Mannerbro, R., Ranlöf, M., Robinson, N., and Forchheimer, R. (2008) Inkjet printed electrochemical organic electronics. *Synth. Met.*, **158**, 556–560.

152. Gamota, Daniel., Brazis, P., Kalyanasundaram, K., and Zhang, J. (2004) in *Printed Organic And Molecular Electronics* (eds D. Gamota, P. Brazis, K. Kalyanasundaram, and J. Zhang), Kluwer Academic Publishers, Dordrecht, pp. 290–293.

153. Liu, Y., Cui, T., and Varahramyan, K. (2003) All-polymer capacitor fabricated with inkjet printing technique. *Solid-State Electron.*, **47**, 1543–1548.

154. Chen, B., Cui, T., Liu, Y., and Varahramyan, K. (2003) All-polymer RC filter circuits fabricated with inkjet printing technology. *Solid-State Electron.*, **47**, 841–847.

155. Pede, D., Serra, G., and De Rossi, D. (1998) Microfabrication of conducting polymer devices by ink-jet stereolithography *Mater. Sci. Eng.*, *C*, **5**, 289–291.

156. Yoshioka, Y. and Jabbour, G.E. (2006) Desktop inkjet printer as a tool to print conducting polymers. *Synth. Met.*, **156**, 779–783.

157. Ngamna, O., Morrin, A., Killard, A.J., Moulton, S.E., Smyth, M.R., and Wallace, G.G. (2003) Inkjet printable polyaniline nanoformulations. *Langmuir*, **23**, 8569–8574.

158. Winther-Jensen, B., Clark, N., Subramanian, P., Helmer, R., Ashraf, S., Wallace, G., Spiccia, L., and MacFarlane, D. (2007) Application of polypyrrole to flexible substrates. *J. Appl. Polym. Sci.*, **104**, 3938–3947.

159. Mire, C.A., Agrawal, A., Wallace, G.G., Calvert, P., and in het Panhuis, M. (2011) Inkjet and extrusion printing of conducting poly(3,4-ethylenedioxythiophene) tracks on and embedded in biopolymer materials. *J. Mater. Chem.*, **21**, 2671–2678.

160. Piner, R.D., Zhu, J., Xu, F., Hong, S., and Mirkin, C.A. (1999) ''Dip-Pen'' nanolithography. *Science*, **283**, 661–663.

161. Huo, F., Zheng, Z., Zheng, G., Giam, L.R., Zhang, H., and Mirkin, C.A. (2008) Polymer pen lithography. *Science*, **321**, 1658–1660.

162. Wang, W.M., LeMieux, M.C., Selvarasah, S., Dokmeci, M.R., and Bao, Z. (2009) Dip-pen nanolithography of electrical contacts to single-walled carbon nanotubes. *ACS Nano*, **3**, 3543–3551.

163. Hung, S.C., Nafday, O.A., Haaheim, J.R., Ren, F., Chi, G.C., and Pearton, S.J. (2010) Dip pen nanolithography of conductive silver traces. *J. Phys. Chem. C*, **114**, 9672–9677.

164. Senesi, A.J., Rozkiewicz, D.I., Reinhoudt, D.N., and Mirkin, C.A. (2009) Agarose-assisted dip-pen nanolithography of oligonucleotides and proteins. *ACS Nano*, **3**, 2394–2402.

165. Lu, H.H., Lin, C.Y., Hsiao, T.C., Ho, K.C., Tunney, J., Yang, D., Evoy, S., Lee, C.K., and Lin, C.W. (2007) Nano fabrication of conducting polymers for NO gas by Dip pen nanolithography. *Conf. Proc. IEEE Eng. Med. Biol. Soc.*, **2007**, 2253–2256.

166. Lim, J.H. and Mirkin, C.A. (2002) Electrostatically driven dip-pen nanolithography of conducting polymers. *Adv. Mater.*, **14**, 1474–1477.

167. Su, M., Aslam, M., Fu, L., Wu, N., and Dravid, V.P. (2004) Dip-pen

nanopatterning of photosensitive conducting polymer using a monomer ink. *Appl. Phys. Lett.*, **84**, 4200–4202.

168. Maynor, B.W., Filocamo, S.F., Grinstaff, M.W., and Liu, J. (2001) Direct-writing of polymer nanostructures:? Poly(thiophene) nanowires on semiconducting and insulating surfaces. *J. Am. Chem. Soc.*, **124**, 522–523.

6
Organic Bionics – Where Are We? Where Do We Go Now?

The advent of novel organic conductors in recent years, in particular nanostructured carbons and conducting polymers, provides us with unprecedented possibilities in terms of enabling more effective communications between biology and electronics. We are on the verge of an exciting new era in the development and application of a wide range of novel bionic devices based on these new materials. Here, we summarize the current state of development and explore the way forward, particularly highlighting the challenges that lie ahead.

Organic conducting materials provide an ideal platform for bridging the gap between biology and electronics. The chemistries available that enable precise attachment of active bioentities to carbon structures leads to the creation of sophisticated interfaces for controlling cellular function and electron transfer processes. The ability to tune electronic and mechanical properties via nanostructuring carbons also presents exciting opportunities.

Organic conducting polymers bring further dimensions to the bionic materials inventory. The ability to control the immediate chemical environment and hence the material–cellular interface, as well as to tune electronic and mechanical properties *in situ*, provides unprecedented real-time control. As with carbon, our excursion into the nanodomain has opened up new routes to tune the mechanical and electronic properties of organic conducting polymers.

Numerous *in vitro* studies have shown the benefits of using these organic conductors to communicate with living cells. These include studies on nerve [1, 2], muscle [3], and bone [4], where the use of electrical stimulation to facilitate cell growth is implemented with a view to assisting regeneration processes.

In vivo studies are, as expected, less prevalent but highly significant. For example, George *et al.* [5] initially examined the *in vitro* biocompatibility of polypyrrole (PPy) using dissociated primary cerebral cortical cells. They used varying electrodeposition synthesis conditions to produce different surface properties with neural networks successfully growing on all of the PPy surfaces. They then surgically implanted PPy implants into the cerebral cortex of the rat and compared the results to stab wounds and Teflon

Organic Bionics, First Edition. Gordon G. Wallace, Simon E. Moulton, Robert M.I. Kapsa, and Michael J. Higgins.
© 2012 Wiley-VCH Verlag GmbH & Co. KGaA. Published 2012 by Wiley-VCH Verlag GmbH & Co. KGaA.

implants of the same size. Quantification of the intensity and extent of gliosis at three- and six-week time points demonstrated that all versions of PPy were at least as biocompatible as Teflon and in fact performed better in most cases. In all of the PPy implant cases, neurons and glial cells enveloped the implant. In several cases, neural tissue was present in the lumen of the implants, allowing contact of the brain parenchyma through the implants.

In a more recent study, Asplund et al. [6] implanted the organic conducting polymer poly(3,4-ethylenedioxythiophene) (PEDOT) doped with heparin into rodent cortex tissue. No cytotoxic response was seen to any of the PEDOT materials tested, although examination of cortical tissue exposed to polymer-coated implants showed extensive glial scarring around implanted platinum or conducting polymer-coated platinum. However, quantification of immunological responses by way of distance measurements from the implant site to the closest neuron, in addition to counting of macrophages, microglia, and cell density around the implant, was comparable to those of platinum controls. These results indicated that PEDOT:heparin surfaces were noncytotoxic and show no marked difference in immunological response in cortical tissue compared to pure platinum controls.

These studies by George et al. [5] and Asplund et al. [6], as well as many others, clearly demonstrate the attractive prospect of utilizing these new organic conductors in bionic applications. An exciting outlook for the future is demonstrated through the recent use of flexible microelectrodes comprising organic conducting polymers [7]. Blau et al. [7] demonstrated that organic conducting polymers (in this case PEDOT doped with polystyrene sulfonate (PSS)) could replace traditional conductors used in microelectrode electrophysiology, a widespread technique used for the extracellular recording of bioelectrical signals. The bendable, somewhat stretchable, noncytotoxic, and biostable all-polymer microelectrode arrays with up to 60 electrodes reliably captured action potentials and local field potentials from acute preparations of heart muscle cells and retinal whole mounts and in vivo epicortical and epidural recordings, as well as during the long-term in vitro recordings from cortico-hippocampal cocultures.

Implanting electrical devices in the nervous system to treat neural diseases is becoming more common and the success of these brain–machine interfaces is critically dependent on the electrodes that come into contact with the neural tissue. Carbon nanotubes are making headway by enhancing neural recording and stimulation in cutting edge in vivo studies. A US-based research team [8] showed that conventional tungsten and stainless steel wire electrodes could be coated with carbon nanotubes using electrochemical techniques under ambient conditions. The carbon nanotube coating enhanced both recording and electrical stimulation of neurons in culture, rats, and monkeys by decreasing the electrode impedance and increasing charge transfer. This study is pivotal to the advancement of implementing organic bionic implants into humans as it is one of only a very small number of studies performed on primates.

These and many other studies to date have stoked the fire with the excitement usually found in an emerging field. However, much has to be done to realize clinically relevant devices utilizing organic conductors.

There is an awareness that to realize real clinical applications the often isolated areas of material design and selection, material synthesis/processing, characterization, device fabrication, and clinical studies must become more integrated.

6.1
Materials Design and Selection

Developing computer models that predict the behavior of more conventional engineering materials in operational environments has revolutionized those areas of materials science.

Computer modeling of protein interactions on nanostructured surfaces has gained recent attention [9], but studies are limited by the computing power available to deal with these complex dynamic systems. The level of complexity enters several new dimensions when we introduce dynamic structures such as organic conducting polymers – when we think of predicting how these might behave in a living biological system, it is beyond the scope of the rather simple modeling approaches available today. The development of more appropriate models will depend critically on our experimental findings in the area of organic bionics in the next five years. So, it appears that at least new modeling approaches will, at best, provide a very rough guide to materials selection.

For regenerative bionics, a further challenge exists in that we must continue to develop biodegradable organic conductors that retain all of the other physical, chemical, and biological properties discussed throughout this text. Ideally, the biodegradation of a regenerative bionic device will be coupled to the healing process. A biological trigger to switch on the degradation process at an appropriate point in time or the use of an externally derived stimulus to promote the trigger for a cascade of events is desirable. Of course, we must pay attention to the nature of the degradation products. They must not cause biological damage, and ideally should assist the regeneration process. The degradable bionic devices should perform their function before being broken down by the body's natural processes and then excreted.

So, for now at least we must learn by experiment, but there must be scope for streamlining the selection experiments required. This will require a more clearly defined relationship between *in vitro* and *in vivo* results (will need to be clear for each clinical application at hand) so that simple *in vitro* experiments can de designed. This is an area that requires the input of great minds to realize significant outcomes in one research life.

6.2
Materials Synthesis and Processing

Most of the materials discussed in this manuscript present a particular challenge to the conventional synthetic chemist and chemical engineer. In our quest to exploit the fascinating properties of nanostructured carbons and organic conducting polymers as quickly as possible, we have often not paid due respect to the optimization of the synthesis and processing conditions needed to achieve optimal and robustly reproducible properties. When that attention to detail is paid, the returns in terms of reproducible materials can be achieved.

It has often been said that the synthesis of conducting polymers is not reproducible – NO, that is not correct. Conducting polymers are often not reproducibly synthesized is a true observation; a subtle but important difference! So, as with all organic synthesis, care must be taken to ensure the purity of the starting materials, control of the reaction conditions, and effective collection/purification of the product.

Similar difficulties have been encountered in the area of nanostructured carbons with some added complexity. With carbon nanotubes or graphene, it is very difficult to provide a material that is homogeneous in either the chemical or the physical domain. The reality is that materials made from nanostructured carbons will be a mixture of different molecular species and/or components of varying dimensions. These variations will result in a distribution of physical, chemical, and biological properties throughout material structures containing these organic conducting components.

We must continue to develop materials processing methods so that organic conducting materials with exquisite control over molecular weight, composition, and dimensions are produced at a scale required for efficient device fabrication.

6.3
Flexible and Printable Electronics

This area of science and technology has progressed at an astounding rate over the past decade [10]. New applications are expected to emerge from stretchable, large-area electronics, particularly with devices such as organic field effect transistors that now have the ability to detect pressure and temperature stimuli [11]. Even when they are stretched by up to 25%, the whole electronic system is fully functional, a feat that is likely to see their use in a variety of different configurations and environments. Elastic electronic components have the potential to elevate new paradigms in flexible robotics [12] and medical tools for steering implants [13]. To further improve their stretchability, the integration of printable organic conductors, which have their own natural elasticity, onto mechanically flexible substrates has a positive impact in the flexible electronics field [14]. There is no doubt here that the development of novel solution-processible organic, metal, and composite conductors will be of importance [15].

The work of Malliaras [16] and others using organic electrochemical transistors to monitor electrical activity from cells gives us an inkling of the future. Not only can organic conductors provide the features highlighted throughout this text but they can also be configured in such a way to produce electronic effects normally associated with inorganic (e.g., silicon-based) materials. The design of hybrid systems, wherein the electronic circuitry based on organic conductors is effectively integrated with biological systems, is realistic and starts to change how we think about medical bionic devices.

6.3.1
Power Supply

Each new bionic application raises special requirements with regard to power supply. While the cardiac pacemaker provides effective operation through an implantable battery, this would not be appropriate for advances in *regenerative bionics* where a biodegradable system is required. The device might ideally have an inbuilt power supply, although the supply of energy via RF (radiofrequency) communication will be a viable option for some applications.

Implantable fuel cells offer some exciting possibilities for meeting the implanted device's power requirements. In particular, enzymatic biofuels are finding their niche in the implantable medical device area such as miniature autonomous sensor–transmitter packages [17]. As a result, there is an active research area dealing with the miniaturization of enzymatic biofuel cells with significant advances being reported by Heller and Mano [18, 19]. However, there is much to be done in developing the next-generation implantable power supply systems.

6.4
Characterization

6.4.1
Characterization Tools

In vitro studies to date have highlighted the challenges that exist in characterizing the inherently dynamic character of material–cellular interfaces. It is widely recognized that spatiotemporal resolution is required to ensure effective bionic communications. The development of tools that enable us to map and visualize molecular, mechanical, and electrochemical events during electronic communication is no trivial task. The advent of biological-atomic force microscopy (Bio-AFM) – and related nanoscale probing techniques presents some new and valuable insights [20]. Mapping with probes capable of interrogating topographical, mechanical, electronic, and chemical information in the nanoscale domain is uniquely possible.

For example, nanoscale electrodes integrated with AFM probes of approximately <50 nm in radius are capable of mapping enzyme activity by detecting enzymatic products while simultaneously imaging the nanoscale surface topography [21]. Such multifunctional nanoprobes are exquisite for exploring the local environment of living cells in which case a probe located in the near vicinity of the cell membrane detects chemical neurotransmitters [22], products of oxidative stress [23], or expression of protein receptors via chemical recognition forces [24]. The use of the probe extends beyond just recording purposes, as its lateral positioning with nanometer precision enables the delivery of physical, chemical, and electrical stimuli to subcellular structures of single living cells. The stimuli can be in the form of mechanical compressive forces to induce calcium signaling [25] or targeted delivery of biomolecules such as DNA sequences via direct insertion of the functionalized probe into the cell [26]. Furthermore, the cellular response is monitored using optical/fluorescence techniques, which combine with AFM and other probing techniques to form a new breed of hybrid characterization tools. In many of the cellular-material interactions highlighted in chapter 4, it is apparent that the inextricable linkage of the physical and chemical heterogeneous surface properties (e.g., roughness, modulus, surface charge, and biomolecule distribution) imparted by the entrapped biomolecules makes it difficult to isolate the cause of cell–substrate interactions. Future characterization of the composites at the nanoscale using techniques such as AFM look promising for elucidating the fundamental interaction mechanisms.

6.4.2
Device Fabrication

To be practically utilized, the organic conductors described here must be effectively integrated into 3D structures with appropriate spatial distribution of incorporated biological constituents over both cellular and molecular dimensions. Temporal control over the accessibility/activity of these constituents and over the electrical stimulation paradigms to be used is also a desired feature.

A combination of electrical, spatial, and temporal requirements introduces extra dimensions that need addressing under nonconventional and adaptive device fabrication strategies. Both fiber (nano- and micrometer-size) spinning and 3D printing techniques provide possible ways forward. A challenge remains for both fiber and printing techniques in the direct incorporation of living cells within the polymer support structures that are intimately associated with organic conductors for bionic devices. The importance of the fabrication of polymer structures with controlled microdimensions in relation to the development of advanced platforms for cell culturing and tissue engineering has been discussed in this book. A major consideration in this regard is the nature of the tissue to be repaired or engineered, and the potential of the polymer structure to promote appropriate growth of cells that best reflect the structural and functional requirements of the tissue. For example, substrates containing aligned fibers spun from biocompatible polymers have been shown to be effective in guiding the growth of several cell types [27–29], including muscle [30]. The formation of fibers also facilitates their fabrication into more complex 3D structures through the implementation of knitting and braiding techniques.

Three-dimensional bioprinting utilizes three-dimensional delivery devices for the rapid and accurate placement of biological materials into biocompatible environments, where postprinting self-assembly takes place. The advantage of 3D printing is that it can be used to create complex spatial patterns directly. The goal of tissue engineering research that aims to utilize 3D scaffold fabrication techniques is to codeposit cells with the supporting matrix during the 3D scaffold fabrication by means of a trivial process, as described by Mironov *et al.* [31]. Mironov *et al.* [32] showed how developmental biology can be applied to organ printing and went on to describe the essential steps and elements of this novel technology. In addition, they discussed the challenging technological barriers and the possible strategies to overcome them, and estimated the overall feasibility of printing 3D human tissues and organs. More recently, extrusion printing was used to construct functional living structures of prescribed shape from a range of cell types [33]. The approach taken mimicked early morphogenesis and is based on the realization that the genetic control of developmental patterning through self-assembly involves physical mechanisms. Three-dimensional tissue structures were formed through the postprinting fusion of the bio-ink particles, in analogy with early structure-forming processes in the embryo that utilize the apparent liquid-like behavior of tissues composed of motile and adhesive cells [33].

It is already obvious that new biofabrication machinery must evolve in parallel with advances in material compositions and configurations if effective progress toward clinically relevant devices is to be made in a realistic time frame.

6.4.3
Pulling It Together . . .

All too often, advances in materials science are compartmentalized into

- material design/synthesis,
- processing,

- device fabrication, and
- characterization.

This is a rather inefficient approach and in the case of organic bionics, wherein progress in any one area is interdependent on the other, it would be entirely ineffective. There is a need for integration of such activities from a technical point of view.

There is a need for parallel, integrated, and synergistic advances in each of the above materials related areas.

References

1. Chen, S.J., Wang, D.Y., Wang, X., Zhang, P.Y., and Gu, X. (2000) Template synthesis of the polypyrrole tube and its bridging for sciatic nerve regeneration. *J. Mater. Sci. Lett.*, **19** (23), 2157–2159.

2. Richardson, R., Thompson, B.C., Moulton, S.E., Cameron, A., Wallace, G.G., Kapsa, R., Clark, G.M., and O'Leary, S. (2007) Polypyrrole with incorporated NT3 promotes auditory nerve survival and neurite outgrowth. *Biomaterials*, **28**, 513–523.

3. Jun, I., Jeong, S., and Shin, H. (2009) The stimulation of myoblast differentiation by electrically conductive sub-micron fibers. *Biomaterials*, **30**, 2038–2047.

4. Saito, N., Usui, Y., Aoki, K., Narita, N., Shimizu, M., Ogiwara, N., Nakamura, K., Ishigaki, N., Kato, H., Taruta, S. *et al.* (2008) Carbon nanotubes for biomaterials in contact with bone. *Curr. Med. Chem.*, **15**, 523–527.

5. George, P.M., Lyckman, A.W., LaVan, D.A., Hegde, A., Leung, Y., Avasare, R., Testa, C., Alexander, P.M., Langer, R., and Sur, M. (2005) Fabrication and biocompatibility of polypyrrole implants suitable for neural prosthetics. *Biomaterials*, **26** (17), 3511–3519.

6. Asplund, M., Thaning, E., Lundberg, J., Sandberg-Nordqvist, A.C., Kostyszyn, B., Inganas, O., and von Holst, H. (2009) Toxicity evaluation of PEDOT/biomolecular composites intended for neural communication electrodes. *Biomed. Mater.*, **4**, 045009–045012.

7. Blau, A., Murr, A., Wolff, S., Sernagor, E., Medini, P., Iurilli, G., Ziegler, C., and Benfenati, F. (201) Flexible, all-polymer microelectrode arrays for the capture of cardiac and neuronal signals. *Biomaterials*, **32**, 1778–1786.

8. Keefer, E.W., Botterman, B.R., Romero, M.I., Rossi, A.F., and Gross, G.W. (2008) Carbon nanotube coating improves neuronal recordings. *Nat. Nanotechnol.*, **3**, 434–439.

9. Cristea, P.D., Tuduce, R., Arsene, O., Dinca, A., Fulga, F., and Nicolau, D.V. (2011) Modeling of biological nanostructured surfaces, in *Nanoscale Imaging, Sensing, and Actuation for Biomedical Applications VII* (eds A.N. Cartwright and D.V. Nicolau), SPIE-International Society Optical Engineering, Bellingham.

10. Forrest, S.R. (2004) The path to ubiquitous and low-cost organic electronic appliances on plastic. *Nature*, **428**, 911–918.

11. Sekitani, T. and Someya, T. (2010) Stretchable, large-area organic electronics. *Adv. Mater.*, **22**, 2228–2246.

12. Aron, M., Haber, G.P., Desai, M.M., and Gill, I.S. (2007) Flexible robotics. *Curr. Opin. Urol.*, **17**, 151–155.

13. Shoa, T., Madden, J.D., Munce, N.R., and Yang, V.X.D. (2009) Steerable catheters, in *Biomedical Applications of Electroactive Polymer Actuators*

(eds F. Carpi and E. Smela), John Wiley & Sons, Ltd, Chichester, DOI:10.1002/9780470744697.ch11.

14. Das, R.N., Lin, H.T., Lauffer, J.M., and Markovich, V.R. (2011) Printable electronics: towards materials development and device fabrication. *Circuit World*, **37**, 38–45.

15. Arias, A.C., MacKenzie, J.D., McCulloch, I., Rivnay, J., and Salleo, A. (2010) Materials and applications for large area electronics: solution-based approaches. *Chem. Rev.*, **110**, 3–24.

16. Yang, S.Y., Kim, B.N., Zakhidov, A.A., Taylor, P.G., Lee, J.-K., Ober, C.K., Lindau, M., and Malliaras, G.G. (2011) Detection of transmitter release from single living cells using conducting polymer microelectrodes. *Adv. Mater.*, **23**, H184–188.

17. Heller, A. (2004) Miniature biofuel cells. *Phys. Chem. Chem. Phys.*, **6**, 209–216.

18. Mano, N. and Heller, A. (2003) A miniature membraneless biofuel cell operating at 0.36 V under physiological conditions. *J. Electrochem. Soc.*, **150**, A1136–A1138.

19. Mano, N., Mao, F., and Heller, A. (2004) A miniature membrane-less biofuel cell operating at +0.60 V under physiological conditions. *ChemBioChem*, **5**, 1703–1705.

20. Muller, D.J. and Dufrene, Y.F. (2008) Atomic force microscopy as a multifunctional molecular toolbox in nanobiotechnology. *Nat. Nanotech.*, **3**, 261–269.

21. Kranz, C., Kueng, A., Lugstein, A., Bertagnolli, E., and Mizaikoff, B. (2004) Mapping of enzyme activity by detection of enzymatic products during AFM imaging with integrated SECM-AFM probes. *Ultramicroscopy*, **100**, 127–134.

22. Cheng, J.K., Wang, W., Wu, W.Z., Huang, W.H., and Wang, Z.L. (2008) Probing brain chemistry – monitoring of chemical signal molecules release from single-cell, single-vesicle, synaptic cleft and morphological analysis with

nanoelectrochemistry. *Chem. J. Chin. Univ.-Chin.*, **29**, 2609–2617.

23. Zhao, X.C., Diakowski, P.M., and Ding, Z.F. (2010) Deconvoluting topography and spatial physiological activity of live macrophage cells by scanning electrochemical microscopy in constant-distance mode. *Anal. Chem.* **82**, 8371–8373.

24. Duman, M., Pfleger, M., Zhu, R., Rankl, C., Chtcheglova, L.A., Neundlinger, I., Bozna, B.L., Mayer, B., Salio, M., Shepherd, D. *et al.* (2010) Improved localization of cellular membrane receptors using combined fluorescence microscopy and simultaneous topography and recognition imaging. *Nanotechnology*, **21**, 115504.http://dx.doi.org/10.1088/0957-4484/21/11/115504

25. Charras, G.T. and Horton, M.A. (2002) Single cell mechanotransduction and its modulation analyzed by atomic force microscope indentation. *Biophys. J.*, **82**, 2970–2981.

26. Cuerrier, C.M., Lebel, R., and Grandbois, M. (2007) Single cell transfection using plasmid decorated AFM probes. *Biochem. Biophys. Res. Commun.*, **355**, 632–636.

27. Bhattarai, S.R., Bhattarai, N., Yi, H.K., Hwang, P.H., Cha, D.I., and Kim, H.Y. (2004) Novel biodegradable electrospun membrane: scaffold for tissue engineering. *Biomaterials*, **25**, 2595–2602.

28. Yoshimoto, H., Shin, Y.M., Terai, H., and Vacanti, J.P. (2003) A biodegradable nanofiber scaffold by electrospinning and its potential for bone tissue engineering. *Biomaterials*, **24**, 2077–2082.

29. Patel, S., Kurpinski, K., Quigley, R., Gao, H., Hsiao, B.S., Poo, M.M., and Li, S. (2007) Bioactive nanofibers: synergistic effects of nanotopography and chemical signaling on cell guidance. *Nano Lett.*, **7**, 2122–2128.

30. Huang, N.F., Patel, S., Thakar, R.G., Wu, J., Hsiao, B.S., Chu, B., Lee, R.J., and Li, S. (2006) Myotube

assembly on nanofibrous and micropatterned polymers. *Nano Lett.*, **6**, 537–542.

31. Mironov, V., Prestwich, G., and Forgacs, G. (2007) Bioprinting living structures. *J. Mater. Chem.*, **17**, 2054–2060.

32. Mironov, V., Boland, T., Trusk, T., Forgacs, G., and Markwald, R.R. (2003) Organ printing: computer-aided jet-based 3D

tissue engineering. *Trends Biotechnol.*, **21** (4), 157–161.

33. Jakab, K., Norotte, C., Damon, B., Marga, F., Neagu, A., Besch-Williford, C.L., Kachurin, A., Church, K.H., Park, H., Mironov, V. *et al.* (2008) Tissue engineering by self-assembly of cells printed into topologically defined structures. *Tissue Eng. A.*, **14** (3), 413–421.

Index

Organic Bionics, First Edition. Gordon G. Wallace, Simon E. Moulton, Robert M.I. Kapsa,
and Michael J. Higgins.
© 2012 Wiley-VCH Verlag GmbH & Co. KGaA. Published 2012 by Wiley-VCH Verlag GmbH & Co. KGaA.